MATHEMATICAL CONCEPTS AND METHODS IN SCIENCE AND ENGINEERING

Series Editor: **Angelo Miele**
Mechanical Engineering and Mathematical Sciences
Rice University

A Continuation Order Plan is available for this series. A continuation order will bring delivery of each new volume immediately upon publication. Volumes are billed only upon actual shipment. For further information please contact the publisher.

Applied Mathematics
An Intellectual Orientation

Applied Mathematics
An Intellectual Orientation

Francis J. Murray

Duke University
Durham, North Carolina

PLENUM PRESS · NEW YORK AND LONDON

Library of Congress Cataloging in Publication Data

Murray, Francis Joseph, 1911-
  Applied mathematics.
  (Mathematical concepts and methods in science and engineering; v. 12)
  Includes index.
  1. Mathematics—1961-   I. Title.
QA37.2.M87   510   78-9012
ISBN 0-306-39252-6

© 1978 Plenum Press, New York
A Division of Plenum Publishing Corporation
227 West 17th Street, New York, N.Y. 10011

Printed in the United States of America

standing extends beyond recorded history and is fascinating in itself. If this development is ignored, technical competence lacks a vital intellectual dimension.

The student may also find it difficult to appreciate that mathematics is something we do, not merely something we know. Theoretical understanding requires complementary mathematical procedures in order to provide guidance for our actions. Originally mathematics was purely algorithmic. The texts of ancient arithmetics simply described how to get the answer. We shall see that ancient geometry was essentially constructive and that this is also true of pure mathematics in the modern sense. The constructive viewpoint of mathematics is unifying and counteracts the pedagogical division into courses.

On the practical level, the applied mathematician must always approach his problem in a quantitative manner. Since he is part of a cooperative effort, he must carefully document all his activities. He must concern himself with all relevant circumstances of his problem and its historical development. The capability and desire to see a situation as a whole is an essential intellectual characteristic of the applied mathematician.

1.3. Opportunities in Applied Mathematics

The Federal Government is, of course, the biggest source of vocational opportunities in applied mathematics, although there are quite a number of corporation laboratories associated with special service areas such as communication or transportation that may also represent vocational opportunities, especially in relation to large-scale automatic computation.

An astronaut has been quoted as saying that when one goes through the prelaunch procedure above a million pounds of highly energetic and potentially explosive fuel, it is fascinating to consider that every item in the system was obtained from the lowest bidder. The Federal Government has only limited production facilities, confined to printing and certain military and naval items, and most of the enormous amount of supplies and equipment it requires is procured by contract. The larger part of the research facilities in the country also are not owned by the Federal Government. Therefore, the bulk of research and development needed for federal operations is accomplished by contractors, but the government does require laboratories and specialized capabilities to oversee this work. Thus the applied mathematician may work on a federally funded project either as part of a contractor's effort or in a government laboratory.

Research and development is a reasonably stabilized part of federal activity, corresponding to about 6% of the total budget. Tables 1 and 2 show

Table 1. Expenditure for Research and Development[a]

	Expenditures (millions of dollars)						
Agency	Fiscal 1970	1971	1972	1973	1974	1975	1976 est.
Defense	7,424	7,541	8,117	8,417	8,791	9,189	9,468
NASA	3,699	3,337	3,373	3,271	3,281	3,181	3,406
HEW	1,235	1,288	1,513	1,791	1,888	1,862	2,423
AEC	1,346	1,303	1,298	1,361			
ERDA					1,475	1,862	2,423
NSF	293	335	418	428	571	571	602
Agriculture	288	315	349	349	377	418	486
Transportation	246	198	274	312	328	307	338
Interior	153	175	210	235	202	265	307
EPA	38	101	133	145	163	207	324
Commerce	118	114	165	179	177	220	239
VA	58	61	66	75	80	97	99
HUD	14	9	47	48	58	52	57
Justice	5	22	13	24	44	44	50
Others	180	154	127	150	185	179	230
Research	5,506	5,685	6,169	6,428	6,783	6,355	7,192
Development	9.592	9.321	9.934	10,356	10,739	12,344	13,199
Total	15,098	15,005	16,103	16,784	17,522	18,699	20,391

[a]*Source: Special Analysis, Budget of United States Government,* Office of Management and Budget. Fiscal 1977, Study P; 1976, Study P; 1975, O; 1974, P; 1973, R; 1972, R; 1971, Q.

the expenditures in millions of dollars in various departments. Table 3 indicates the variation over a longer period.

A major part of the expenditure for research and development is in the Department of Defense. There is a continuous development of missile weapon systems, military aircraft, submarines and antisubmarine systems, land-combat weapon systems, and electronic devices used in warfare. This development tries to exploit all available technology and requires continuous engineering development, involving simulations with mathematical, logical, and computational aspects. Simulations are also used in military planning and training.

The next largest area of research and development is in the National Aeronautics and Space Administration with its emphasis on the exploration of space and the development of means for this purpose. Air traffic control also requires systems with sensing and computational abilities. The Department of Health, Education and Welfare is immediately concerned with biological research but it also needs large-scale information storage and retrieval systems. The Atomic Energy Commission dealt with nuclear weapon systems

Table 2. Current Expenditures for Research and Development[a]

| | Expenditures (millions of dollars) | | | |
Agency	Fiscal 1976	Fiscal 1977	1978 est.	1979 est.
Defense	9,329	10,176	11,137	12,315
Energy		3,181	3,881	4,188
ERDA	2,225			
NASA	3,521	3,763	3,824	4,090
HEW	2,566	2,591	2,890	3,141
NSF	623	650	715	764
Agriculture	460	516	604	582
EPA	251	372	330	345
Transportation	303	311	342	335
Interior	315	293	335	342
Commerce	224	233	270	298
Nuclear Regulatory Commission	81	104	125	145
VA	97	105	112	113
HUD	54	66	54	54
TVA	19	25	36	52
AID	23	38	35	48
Others	142	141	165	163
Total	20,233	22,462	24,854	26,984

[a]Source is the same as that of Table 1, Fiscal 1978, Fiscal 1979, Study P.

and nuclear power systems, and these require engineering with a large mathematical component. The energy aspects have been taken over by ERDA, which has a somewhat more general responsibility in regard to sources. The government is also involved in the planning of air and ground systems of transportation and developing more completely mathematical methods of weather prediction. Simulations are also of considerable importance to the Environmental Protection Agency.

The ability of a modern data-processing system to deal with an enormous amount of information constitutes a challenge to tackle new classes of problems. One example is the quantitative description of the national economic system in terms of the exchange of goods and services. Another example would be an information storage system with a flexible retrieval system to deal with what is known concerning chemical compounds that can be manufactured. Many of these essentially large problems are of national interest and hence of concern to the Federal Government; for example, the Department of Defense may want to determine the economic capability of the country to sustain a certain war effort. One would anticipate that the

Table 3. Historical Summary of Expenditures for Research and Development [a]

Fiscal year	Expenditures (millions of dollars)					
	DOD	NASA	AEC→ERDA	HEW	NSF	Total
1954	2,487	90	383	63	4	3,148
1955	2,630	74	385	70	9	3,308
1956	2,639	71	474	86	15	3,446
1957	3,371	76	657	144	31	4,462
1958	3,664	89	804	180	34	4,991
1959	4,183	145	877	253	54	5,806
1960	5,654	401	986	324	64	7,744
1961	6,618	742	1,111	374	83	9,284
1962	6,812	1,251	1,284	512	113	10,381
1963	6,844	2,539	1,336	632	153	11,999
1964	7,517	4,171	1,505	793	203	14,707
1965	6,728	5,093	1,520	738	206	14,889
1966	6,735	5,933	1,462	879	241	16,018
1967	7,680	5,426	1,467	1,075	277	16,842
1968	8,164	4,724	1,594	1,283	315	17,030
1969	7,858	4,252	1,654	1,221	342	16,208
1970	7,424	3,699	1,346	1,235	293	15,098
1971	7,541	3,337	1,303	1,288	335	15,005
1972	8,117	3,373	1,298	1,513	418	16,103
1973	8,417	3,271	1,361	1,791	428	16,784
1974	8,791	3,181	1,475	1,888	571	17,522
1975	9,189	3,181	1,862	1,862	571	18,699
1976	9,468	3,406	2,423	2,423	603	20,391

[a]Source is the same as for Table 1 with emphasis on Fiscal 1971, Study Q.

Environmental Protection Agency will concern itself with the overall economic effects of various technological restrictions. Medical research is frequently dependent on chemical structure information, and one may wish to consider all available chemical compounds with a given substructure. The analysis of complex biochemicals is in general a large-scale data-processing task, and effective understanding of the function of biochemicals is probably a larger problem. Modern sensing devices can gather an enormous amount of information, which must be automatically processed in order to produce maps or crop reports. Novel applications tend to require applied mathematicians when there is no adequate professional discipline available and new logical analysis and constructs are required.

The growth of communication and transportation requirements in recent decades has provided considerable incentive to private industry to improve technologies and develop new ones. Power requirements have also

Preface

The primary objective of the course presented here is orientation for those interested in applying mathematics, but the course should also be of value to those interested in mathematical research and teaching or in using mathematics in some other professional context. The course should be suitable for college seniors and graduate students, as well as for college juniors who have had mathematics beyond the basic calculus sequence. Maturity is more significant than any formal prerequisite.

The presentation involves a number of topics that are significant for applied mathematics but that normally do not appear in the curriculum or are depicted from an entirely different point of view. These topics include engineering simulations, the experience patterns of the exact sciences, the conceptual nature of pure mathematics and its relation to applied mathematics, the historical development of mathematics, the associated conceptual aspects of the exact sciences, and the metaphysical implications of mathematical scientific theories. We will associate topics in mathematics with areas of application.

This presentation corresponds to a certain logical structure. But there is an enormous wealth of intellectual development available, and this permits considerable flexibility for the instructor in curricula and emphasis. The prime objective is to encourage the student to contact and utilize this rich heritage. Thus, the student's activity is critical, and it is also critical that this activity be precisely formulated and communicated.

The student's efforts outside the classroom should be mainly devoted to a project of his own choice, which he should develop and report to the class. A student should have at least three opportunities for such reports. See also the comment preceding the exercises of Chapter 1. It is not necessary for the instructor to be expert in the topics chosen. The effort of the instructor to understand the means of the student's presentations can yield excellent training.

The exercises are intended to serve a number of purposes. Reports on the exercises can be used to supplement the project presentations by assigning questions concerning distinctly different areas. They should also assist the student in developing his project either by direct inclusion or by indicating various possibilities. In many situations there is need to precisely formulate the problem to be solved, and this is represented in a number of exercises by some ambiguity, which permits a classroom procedure in which the exercises are considered, without previous preparation, to the point of specifying the objectives and proposing methods of attack.

The summaries of sections given in the table of contents are intended to assist someone who has read the book to locate specific discussions. They are not complete summaries in the usual sense.

The author is deeply grateful to the students who attended the course during its development stages and to Walter Sewell and Y. H. Clifton for comment on the text. He would also like to express his appreciation to Gwynne Moore, who drew the illustrations, and to Mrs. Ann Davis, Mrs. Bonnie Farrell, and Mrs. Anne Tunstall for the preparation of the manuscript.

Francis J. Murray

Contents

3. Understanding and Mathematics

4. Ancient Mathematics

5. Transition and Developments

6. Natural Philosophy

7. Energy

8. Probability

9. The Paradox

1

Introduction

1.1. Vocational Aspects

We are concerned with "what is applied mathematics?" and not "how to apply mathematics." In universities, there are many courses that deal with the latter question, and whenever possible we will take advantage of this to refer to their content. Our initial answer is that applied mathematics is the vocational use of mathematics other than in teaching or mathematical research, and we will explore the intellectual developments that are associated with such vocational uses, with emphasis on the aspects not normally part of the "how to" courses.

There are many vocations in which mathematical procedures form an inherent part, for example, physics, engineering, and actuarial practice. Both the applied mathematician and the teacher of mathematics should be interested in the intellectual basis for this type of mathematical application.

However, a mathematics student may also be interested in applied mathematics as a vocation itself. Our civilization is capable of collective actions on many scales, from that of a small manufacturing operation to national enterprises such as space exploration, highway construction, or food distribution. It is notorious that such collective actions can have both desirable and deleterious effects. An ideal scientific understanding of an experience complex would permit a quantitative prediction of the effects of such actions and possible alternatives. Among the many elements needed to obtain such predictions are mathematical analysis and computation. Although the scientific understanding available in practical cases hardly ever approximates the ideal, there are vocational opportunities for persons with a mathematical background. A mathematics student who is interested should obtain an overall understanding of the nature of these applications and in particular of the role of mathematics, a role more subtle than that usually ascribed to it.

The term "applied mathematics" usually implies mathematical and logical discussions of considerable sophistication. At present, new developments tend to take the form of an enormous amount of elementary mathematics in a complex structure and to reflect the availability of automatic data processing. But the manipulations of classical mathematics are still appropriate, and the applied mathematician must concern himself with both types of procedure. To a considerable extent, classical mathematical analysis has become part of the professional skill of other vocations, for example, electrical engineering and physics. But modern developments tend to go beyond the classical limitations of analysis and require novel logical structures combining manipulation and computation, and it is in these cases that the applied mathematician can contribute.

1.2. Intellectual Attitudes

The divisions of a university faculty into departments corresponding to various disciplines may induce the student to believe that this corresponds to some deep resolution of knowledge and understanding into separate and indeed disparate compartments. Our education tends to produce the annelid, or segmented worm, concept of understanding. For example, a situation may first be analyzed from the point of view of economics. This analysis leads to engineering problems, which in turn produce problems in physics or engineering. The latter then yield problems in mathematics, and these refer finally to computation. Thus groups of specialists can each deal with problems in their own field, nicely isolated from the others.

But this type of resolution is essentially inapplicable even when one makes the simplifying approximations that are usually necessary in practice. The anatomy is invariably far too complex to permit such a dissection; indeed the analogy is even unfair to the worm. But complexity itself is not the only element involved. The historical evolution of the academic disciplines and that of the intellectual formats on which current applications are based are quite disparate, and it is pointless to try to fit the latter to a procrustean bed of scholastic subdivisions.

The technical training and specialized skills of the various academic disciplines may be appropriate for a specific problem. But they must be applied within a framework of general overall understanding and frequently in an intellectual format quite independent of academic predilections. The actual situation is much more interesting and challenging than that which one might, naively, expect. The historical development of quantitative under-produced problems in distribution, the quality of the environment, and fuel availability. Electronic computation has yielded an entirely new industry in the 1950s and 1960s associated with an extraordinary combination of technological developments.

In these cases and in civil and mechanical engineering, the fundamental improvement in technology may be to a considerable extent independent of mathematical considerations, except where more sophisticated physics appears, as in the use of quantum mechanics in semiconductor theory. But automatic data processing has made possible a much more complete analysis of the situations involved in the introduction of new developments, and this analysis is very important to obtain optimal results. Engineering procedures were also expanded by the availability of automatic computation to solve classical dynamics problems and for the data processing of automatic sensors involving radar and infrared. Each of these developments is of vocational significance for mathematics. Also experimentation and testing for reliability has tended to become more complex and to make demands that exceed the classical procedures of statistics.

1.4. Course Objectives

We will develop the concept of vocational applied mathematics and use it to obtain insights into other vocational uses of mathematics. The exercises that conclude this chapter are an integral part of the course. We will concentrate on applied mathematics as a part of engineering effort, and we will consider the general character of such efforts and, in particular, the role of simulations. We will consider the technical structure of such simulations and see that they are part of a format of understanding that has been developed in conjunction with the availability of large-scale automatic computation since World War II.

This understanding is essentially based on mathematical concepts and procedures. Standard philosophical approaches are not really adequate to explain this understanding. The relation of modern mathematics, as it is now taught in graduate school, to applications, even the most abstract scientific theories, can only be explained in terms of historical development. This historical development began with ancient arithmetic and geometry and was a complex evolutionary process in which mathematics took on many diverse forms that have left their imprint on notation and terminology and on the intellectual approach of many who apply mathematics. Our study should provide insight into the practical applications of mathematics, the exact sciences, and pure mathematics itself.

Exercises

Term Project: Presumably each student has a specific reason for being interested in applied mathematics. It is appropriate for him to set up a project dealing with a subject matter of his own choice that he will carry out during the term and present in the form of reports to the class. The first report should describe the subject matter, the reasons for choosing it, and the student's objectives in the project. Subsequent reports could consider (i) relevant organizations, (ii) the role of mathematically based procedures, (iii) simulation structures, and (iv) validation concerns.

In the following set of exercises, the student may concentrate on an area of interest to himself. Reports to the class are very desirable. The student should use the library to make contact with fields involved.

1.1. What kind of mathematics is used in the following fields:
- (a) Surveying
- (b) Navigation
- (c) Classical mechanics and elementary physics
- (d) Electricity and magnetism
- (e) Fluid flow and elasticity
- (f) Thermodynamics
- (g) Quantum mechanics
- (h) Inorganic chemistry
- (i) Organic chemistry
- (j) Astronomy
- (k) Astrophysics
- (l) Relativity
- (m) Astrology
- (n) Keeping book on horses
- (o) Electrical engineering
- (p) Civil engineering
- (q) Mechanical engineering
- (r) Economics
- (s) Linguistics
- (t) Computer science
- (u) Finances
- (v) Agriculture

In each instance, how do you know that your answer is reasonably complete? This question can be subdivided and apportioned to the class.

1.2. Obtain a list of companies that hire mathematicians. Also obtain a list of government agencies that hire mathematicians.

1.3. Obtain a list of references describing mathematical procedures used in applied mathematics and normally not requiring large-scale automatic computation.

1.4. You have a roster of consultants listed by academic disciplines. What would be the appropriate disciplines for the following enterprises?
- (a) A national highway system
- (b) An antimissile missile defense system
- (c) A light-weight personal armor
- (d) A sewage-treatment plant
- (e) A crime information network
- (f) A cancer research institute

1.5. What kind of mathematical tables do you know about? What are they used for?

1.6. Instead of bunting, a baseball player tries to hit the ball sharply into the ground so that it will have a high first bounce. Is this a good idea? Give quantitative estimates of the skills and capabilities required.

1.7. How can a pitcher throw a curve? Estimate quantitatively the various effects desired and the required skills and capabilities.

1.8. In working the exercises of this chapter, how much assistance have you received from the text? Why do cars have batteries?

1.9. What mathematical algorithms are you familiar with? What is each used for?

1.10. Describe the mathematical formulation of the behavior of the following systems in the appropriate environment:
- (a) A missile guidance system
- (b) An aircraft with VTOL capability
- (c) A warship relative to its commander
- (d) A tank, half-track, or truck
- (e) An antisubmarine-warfare naval complex operating under one commander
- (f) A medical examination center for military personnel used for induction, retirement examinations, or eligibility for further or special services
- (g) A multiple-target intercontinental ballistic missile
- (h) A system for testing a medium-range land-based ground-to-ground missile
- (i) A fly-by-Jupiter space mission (or a soft landing on Mars or an atmospheric penetration of Venus)
- (j) A space shuttle

1.11. Professional prudence requires that tissue removed in a surgical operation be examined by a hospital pathologist independent of the surgical team. With associated clinical and personal data, the resulting records would constitute a very extensive file of medical information. Describe a system to centralize this information and the purposes it could serve.

1.12. How can you estimate the time it will take to answer a problem of this set?

1.13. Describe the elements that characterize the quality of the environment in terms of numerical parameters. What kind of mathematics would be used to describe the effects of polluting agents?

1.14. There are two major types of nuclear explosive devices. Describe the mathematical formulation of the action in each case.

1.15. What kinds of mathematics are involved in the design of nuclear reactors? Many interesting and very difficult mathematical problems arise from the effort to utilize nuclear fusion as a power source. Describe these problems.

1.16. How is the traffic situation on a highway system described?

1.17. How is the service of a telephone exchange described mathematically? A long-distance telephone network?

1.18. What is the mathematics associated with classical (lumped parameter) circuit theory? How is this changed to handle integrated circuits?

1.19. A considerable number of diverse procedures are used to analyze the structure of the complex molecules of biochemistry. What is the associated mathematic?

1.20. Biochemical activity is usually described in terms of chains of reactions with "inhibitors" and "activators." How would such a situation be described in mathematical or logical terms? How would separation by membranes be taken into account?

1.21. How can the chemical structure of a compound be described in such a way that it can be stored in a computer? If you have stored a list of such compounds, how could you test whether a given compound is in the list? How could you retrieve all compounds with a given substructure?

1.22. If the country is considered as divided into geographical areas for the purpose of economic analysis, what are the quantitative parameters one would associate with each unit? What mathematics would be appropriate to produce a dynamic model of the economy on this basis?

1.23. How would one describe an economic model for international trade? What is the significance of export, import, balance of trade, rate of exchange? What relation would one expect between these variables? What would be the effect of long-range discrepancies?

1.24. Modern mapping is based on aerial photography. Describe the mathematics involved. In this process, topographical information is obtained as numerical data before it is transposed to the usual chart form. What is the magnitude of the data needed to construct, 2 cm to a kilometer, maps of the entire United States if the charts have a resolution of 1 mm and the height data a resolution of 10 m? How can this data be compressed without loss of information? What approximations may be suitable for further compression and how would these be used?

1.25. What are satellites used for, and what mathematics is involved?

1.26. What is a power network? What is its purpose and how is it described mathematically?

1.27. How is the logical structure of a computer described?

1.28. What is the band theory of electrical conduction for semiconductors, and what are the associated mathematical questions?

1.29. What is "operations research"? To what areas has it been applied?

1.30. The terms "design of experiments" and "analysis of data" can be interpreted in different ways. Discuss these for agricultural experiments, weapon-system testing, and nuclear particle research.

1.31. Modern linguistic theories have introduced mathematical constructs and transformations. Describe these.

1.32. Investing in corporate stocks has various quantitative aspects. Mathematical predictions are clearly desirable. How can these be obtained?

1.33. Betting on the outcome of horse races has various quantitative aspects. Mathematical predictions are clearly desirable. How can these be obtained?

1.34. One presumably mathematical area is called "the numbers." Describe this. Discuss the procedures used by the financial interests that run this business to avoid catastrophes due to coherent betting.

1.35. The term "reliability" is applied in many different circumstances. What are the quantitative aspects of the term relative to the following: (a) light bulbs; (b) integrated circuits; (c) automobiles; (d) airlines; (e) army rifles; (f) screw-making machines; (g) rocket fuel; (h) systems of computer software; (i) dictionaries; (j) scientific theories.

2

Simulations

2.1. Organized Efforts

The work of the applied mathematician is usually part of a relatively large effort representing a contractual obligation of his employer. Examples are the engineering development of weapons, aircraft, naval ships, land vehicles, weapon systems, communication systems, transportation systems, computing systems, service systems, and training devices. Government requirements in the present day also demand extensive studies in which the major aspect is economics, sociology, or biology, as in ecological impact studies. Most work of this type is government oriented.

Such efforts are based on a mathematical formulation of scientific and technical understanding. Large-scale computing permits extensive computation for decision purposes. The correctness of this computation may be the immediate responsibility of the applied mathematician, but this cannot be isolated from an understanding of the total effort. The mathematician must participate in a general team effort, and this requires appropriate communication in the form of reports and documents.

The personnel in such an effort usually have a unifying background of common experience. Thus, the scientific understanding utilized tends to have a specific technical cast based on experience and relevant engineering practice. This affects the choice of mathematical procedure as well as the fact that more general procedures can lead to severe mathematical difficulties.

In addition to these large efforts, the mathematician may also be involved in certain specific statistical procedures, and he may also find educational responsibilities. Statistics may be an integral part of a large effort, but it can also be needed in other immediate developments, and the applied mathematician should be familiar with the theory. New technical developments may introduce mathematics that is novel to the people involved, and instruction may be desirable.

2.2. Staging

Economic considerations are always paramount in research and development efforts. If a project is sponsored by a government agency, the value of the ultimate outcome usually has to be determined. When work is done by a private contractor, the resources available are limited and must be used efficiently. Resources include money; management and technical manpower (especially individual capability); production, laboratory, and computing facilities; support personnel and services; and office space.

Because of the limitations and pressures associated with resources, large efforts, generally, are developed in stages that serve two purposes. Each stage either justifies the continuation of the project into the next stage or indicates a termination. Funding considerations are often more complicated than go or no go decisions. Even when a project is not abandoned, the continuation may be postponed or stretched out, and for this reason the stages themselves are structured into phases to permit resource management.

A government project will usually have an "in-house" first stage. A project can arise because experience indicates a need, or a technical development shows that certain possibilities have become available, or the particular agency involved may be subject to external demands. One can consider the first phase as consisting of a number of studies that result in a definition of the project and a set of objectives or requirements for the outcome. A second phase must study feasibility, and when this is completed a decision must be made to commit funds for a larger effort. The continued larger effort will usually be executed under contract, since the agency does not have the required in-house resources to proceed in any other way. The student should appreciate that this is a carefully maintained element in an economic and political system. The contracts have considerable local economic impact, which is politically valuable to congressmen, who in turn support the overall program of the agency.

The stages in the contract work are in general highly visible and are clearly associated with go or no go decisions and the possibility of delays or stretching out of funding. One would normally expect each stage to be individually contracted for in a sequence such that the next commitment can be approximately controlled. Variations from this procedure may occur due to political pressure or intra-agency reasons. The contracts associated with the later stages are more desirable, and there is pressure for early commitment. When the project involves basic innovations whose feasibility should be carefully established by the staging process, early commitments can be quite unfortunate.

Suppose we are dealing with an engineering development that will lead to the manufacture of a large number of similar units. Clearly this will apply to most weapon systems, aircraft, vehicles, and so forth. An engineering development whose objective is a small number of large or very large units such as naval vessels or, say, satellites, which are large in the sense of technical effort, would have a somewhat different staging. There are also intermediate developments, but it is desirable to be definite, and we will consider the stages for the first type.

The first contract phase can be considered to be a planning stage, to be followed by a prototype stage, a production stage, a fielding stage, a service stage with the important element of maintenance, and a phase-out stage. Such a development requires a sequence of decisions and an evolvement of understanding that will permit progress to the next stage. This understanding must have a scientific and technical basis and is usually embodied in simulations.

To use the available resources efficiently, each stage must be precisely structured in time. Thus the planning stage would consist of a feasibility phase, a product-design phase, a development plan, a prototype production plan, a service production plan, training plans, fielding plans, and service and maintenance plans. The corresponding phases could and normally would overlap.

In regard to feasibility, it is usually desirable to expand the previously available in-house studies and precisely establish the requirements. Relative to requirements, the interests of the government and that of the contractor do not normally coincide. A simulation of the expected use of the system, including the environment, may be significant for feasibility. Essential differences in points of view can be either eliminated by agreeing on the technical base of the simulation or brought into sharp focus to fix responsibility.

In product design, it is usual to set up a simulation of the use of the system in various environments on a computer to test whether a tentative design satisfies the requirements. In such simulations it is possible to modify and adjust the tentative design conveniently and obtain performance and cost operation.

If a system design is available, it is possible to plan the rest of the project completely. But for modern production, a very large amount of interrelated data is required, such as specifications, blueprints, production scheduling, and procurement. A structure for this data may be available, relative to a fixed production facility, but the data itself must be produced in a form that will permit design flexibility. Thus the planning of the production stages involves considerable data generation and processing, and these procedures themselves may be automated and designed. If the production facility has to be developed or restructured, one has further design problems. These developments involve considerable technical understanding and may require simulations. The development of other plans will similarly involve considerable data and structuring of procedures and simulations may be utilized.

2.3. Simulations

A simulation is defined in the dictionary as a pretense or feigning. There is a situation in which the subject of the simulation is supposed to be involved. One contrives an imitation of this situation so that the subject has vicarious experience with it. One possible purpose is deception, and many forms of amusement, from roller coasters to opera, are based on simulations.

There are, however, technical situations in which vicarious experience is valuable. These require that the representation be logically equivalent to the original situation. Such simulations can be used for design purposes so as to anticipate the capabilities and limitations of a proposed system. They can also be used for training purposes to develop the understanding, skills, and proper reactions needed to use complex devices. Simulations may be used for planning or to develop understanding. For example, a logical representation of an explosion may indicate phenomena that can be detected experimentally and either confirm or deny certain hypotheses.

It is clear that there are various aspects common to all these simulations. To the subject of the simulation, the original situation is reproduced symbolically and a time history of a specific imaginary experience with the situation is developed. To produce this, the symbols must be activated by some process logically equivalent to the original situation, and in technical situations this is usually a computation, proceeding automatically in a data processing system.

Consider, for example, a flight trainer, which is a device that enables a pilot to learn how to fly a specific type of airplane. The cockpit of the aircraft is simulated so that the trainee judges the flight from instrument readings and moves the airplane controls. This is, of course, the symbolic representation of the situation. The computation in the computer correspond to the flight determined by the controls, and the various numerical quantities that describe this flight are generated as functions of the time and appear on the instruments. The mathematical procedure that is realized by the computation is called the math model for the airplane.

Certain elements are usually present in most computer-activated simulations. The symbolic representation may involve a considerable amount of equipment, which receives output from the computer and produces input. The math model must be based on a scientific and technical analysis of the original situation. Furthermore the math model must be used to produce a computer program that will yield a time development corresponding to a specific experience. Finally there is usually a requirement for an overall evaluation of the total experience with the simulation.

2.4. Influence Block Diagram and Math Model

The math model must incorporate the scientific and technical understanding of the situation that is simulated. The appropriate technical information and experience must be assembled and documented. The initial analysis of the situation corresponds to the "influence block diagram." One determines the various aspects of the situation that can be specified uniquely in a quantitative manner. Usually each aspect is specified by a number of numerical values, and in a specific experience each such aspect will have a time history, which is given by expressing these values as functions of the time.

A block in the diagram corresponds to such an aspect and its related set of quantities. The math model gives the mathematical basis for determining these quantities as functions of the time. For example, certain of these variables may change continuously and correspond to the solution of a simultaneous system of equations in which time is the independent variable. On the other hand, other variables may take on only a discrete set of values and hence will change abruptly during a simulation. Such a change is called a critical event, and the math model must specify the criteria for critical events. In most cases, a criterion of this sort is given by a change of sign of a function of the continuous variables.

These functional relations correspond to relations between the blocks of the diagram and may be indicated by connections on the diagram. But these connections represent more general relations than the specific mathematics of the math model. A preliminary analysis can perhaps establish these relations as a first step in determining the math model.

Various ways can be used to analyze the situation into aspects corresponding to diagram boxes. It may be possible to consider certain components of a device as rigid bodies. For each such component, one has the variables that describe the motion. These variables satisfy differential equations containing forces. The forces depend on other aspects of the situation, and this dependence corresponds to connections in the block diagram. The actual functional relations for these forces yields the math model. However various forces or sets of forces may themselves correspond to boxes in the diagram. In simulating the flight of an airplane, the aerodynamic forces and torques can be considered such an aspect. These forces depend on the velocity and altitude of the aircraft and the position of the control surfaces.

But analysis may require more sophisticated geometric constructions as, for example, in simulations of explosions, combustion, internal ballistics, and chemical processes. The situation has a number of regimes of activity in time, and during each regime at each instant of time, one has a spatial division into regions and separating surfaces. The different regimes can be numbered, and this number can be considered a "state variable," which changes abruptly at critical events. Within each regime, the substance in a given region may be considered an aspect that has a quantitative description in terms of its motion and thermodynamic characteristics. If the substance is not a rigid body, the motion may be quite complex. A theoretical basis for describing such motions is available in the form of partial differential equations. In general, these partial differential equations have no formal solution. Numerical methods, using automatic data processing, have greatly extended the range of cases which can be approximated by a numerical development in time. However, in these cases, one may need to substitute experimentally determined patterns of motion for theoretical solutions of the equations.

The notion of state variable applies to many simulations. For example, an aircraft can be in different flight states, such as moving on the ground, normal flight, or in a stall or spin. A state variable can also specify whether a certain malfunction obtains. In general, one needs to know the values of all the state variables to determine the equations that describe the change in the continuous variables. Thus, the state variables determine regimes in time during which the continuous variables change in accordance with a given set of rules until a critical event occurs and the set of rules changes. For example, in takeoff, the aircraft moves along the ground with increasing velocity until the aerodynamic lift force exceeds the weight. In flight, the aerodynamic forces will differ from those on the ground and certain forces will disappear. Retracting landing gear corresponds to a change in a state variable.

In general, one would consider each state variable as an independent aspect of the simulated situation to be represented by a box. Other aspects can be identified as having essentially the same significance for different values of the state variables but whose math model may be different. For example, the aerodynamic lift force has a math model that is dependent on the value of the flight state variable. Because of the complexity of the relations of state variables and continuous variables, it is desirable to begin an analysis of the original situation with the construction of the influence block diagram.

The math model gives the change of the continuous variables for each possible combination of values of the state variables. If the continuous variables are specified as solutions of a system of differential equations in which the independent variable is the time, they are said to be given "dynamically." If they are given as functions of time directly, they are said to be given "kinematically." The manner of change may be determined by scientific principles or it may be given by an empirically observed pattern. An example of the latter would be the motion of a ship after a rudder setting has been changed.

It is clear that the determination of the math model on the basis of the influence block diagram, scientific principles, and empirically observed patterns should be carefully documented. When an organization has dealt with similar problems over a period of time, there is usually a traditional procedure for determining the math model but complete documentation is still essential. Most situations will have individual variations.

2.5. Temporal Patterns

The math model specifies a combination of regimes of continuous development followed by critical events, which can be diagrammed as a "flow chart." The influence block diagram and the flow chart represent complementary aspects of the math model, one of which refers to the status at an instance of time, the other refers to a time pattern of behavior.

The influence block diagram and the flow chart are logical consequences of the math model and are determined when the math model is given. But usually one deals with a somewhat more complex situation, in which the math model is not given first. For example, in design procedures, one may have cases in which the given requirements are time patterns that can be incorporated into a flow chart. The objective is to design a system whose math model yields this flow chart. One may also have a procedure based on experience and yielding a design that can be represented by a influence block diagram, but the specific math model may have to be determined. A variation of this situation is one in which a general structure for the design is assumed and this general structure becomes specific when values are assigned to certain parameters called "design parameters." Here it is usual to set up a simulation in which these parameters can be adjusted so as to obtain the desired behavior, which may be represented by a flow chart, or to avoid certain objectionable characteristics.

Thus, while the math model logically determines the influence block diagram and the flow chart and the computational procedures associated with individual experiences, this is not necessarily the order in which they arise in the analysis of simulations. The math model, the influence block diagram, and the flow chart form a logical and symbolic entity that is valuable for engineering and many other situations.

A flow chart can be considered to have a network or linear graph character. We can draw a flow chart as a linear graph in which a regime corresponds to one cells and critical events to end points, or as a network with regimes corresponding to branches and critical events to nodes. But the regimes are directed, of course, and the choice of the new regime that follows a critical event may be determined by something not represented on the flow chart. Nevertheless, any specific time history corresponds to a path on this flow chart. A permissible path of this type is called a "scenario."

These general schematic ideas apply even to simulations that are not engineering in character. An example would be a computerized war game whose purpose is to train staff officers to handle problems of manpower, equipment, petroleum supplies, and other requirements of a military campaign. In the initial planning for such a simulation, a specific time history, or scenario, is determined. During the game the computer must produce a running account of the above military essentials, taking into account the attrition due to the campaign itself and to enemy action and various factors such as terrain, and also taking into account the actions of the trainees. These elements appear in the influence block diagram, and the math model is based on them using statistics acquired from military experience. Notice that here the natural order is scenario, block diagram, and math model.

A war game whose purpose is to test a plan or to develop skill in tactics will have a somewhat different development. The initial effort usually would be concerned with the influence block diagram. The opposing forces would be resolved into groups whose status can be quantitatively described and treated as a unit relative to combat and mobility. These units interact in various ways, such as by fire power, losses, terrain, map position, and logistics. These relations are relatively complex and are perhaps analyzed best initially in diagrammatic form, and the math model is based on this diagram. The limitations of the analysis result in a procedure in which values for manpower, fire power, vehicles in service, ammunition, and supplies are extrapolated for brief intervals of time. The math model must provide the rates for these extrapolations in terms of the variables. The simulation is essentially dynamic.

2.6. Operational Flight Trainer

To illustrate the ideas associated with technical simulations, in particular, block diagram, math model, flow chart, and scenario, we consider an operational flight trainer (OFT). Such a device is used either to train novices to to familiarize experienced pilots with a new aircraft. Our discussion will be simplified to the greatest extent consistent with our purpose.

The very interesting technical history and characteristics of OFTs are described in the U.S. Navy, *Commemorative Technical Volume.*[2] An introduc-

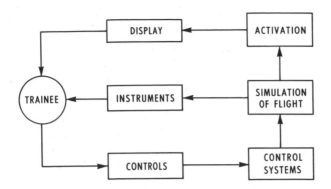

Figure 2.1. Operational flight trainer.

tion to the aerodynamic aspects and equations of motion is given by Connolly.[3]

We suppose that the purpose of the trainer is to provide familiarity and understanding with powered flight and develop skill patterns for novice pilots. The trainer will consist of two portions. One part (see Figure 2.1) is a cockpit with instruments and controls and a display to provide cues needed for takeoff and landing. The other portion of trainer involves data processors that contain a mathematical simulation of the aircraft flight determined by the cockpit controls. This simulation governs the cockpit instruments and provides input for the activation of the display. The servo systems that position the control surfaces on the aircraft are also simulated.

Our immediate concern is with the flight simulation, and this is based on Newtonian mechanics. However, the detailed analysis must be based on the flight phase, since that determines the forces and torques. The normal sequence of flight phases is (1) takeoff one, (2) takeoff two, (3) normal flight, (4) glide, (5) landing one, (6) landing two. However, of these phases, (1), (2), (5), or (6) could end in disaster, for example, if a wing scraped the ground. On the other hand flight or glide could end in a stall, spin, or tumble. But in both the disaster case or the cases of abnormal flight, the simulation must produce appropriate output. The trainee may also be capable of producing a return to normal flight from the abnormal case. We have, therefore, a finite set of possible flight phases and the dynamic analysis is determined by the phase. The simulation in each phase ultimately leads to a change of phase. We can, of course, introduce a discrete variable to specify the flight phase.

Thus the simulation of flight box in the above can be resolved by introducing a flight phase box and a number of subsidiary boxes (Figure 2.2). The significance of the double lines is that while the flight phase determines the analysis to be used, i.e., the subsidiary boxes, the development in time in each

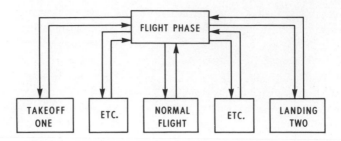

Figure 2.2. Resolution of the flight phase.

case can lead to a change of flight phase. We will discuss this in more detail later.

We will concentrate our attention on the subsidiary box "normal flight." The two outputs of the simulation of flight are the instruments and the display activation. The landing display is determined by the relative position of the aircraft and the landing field. There are two classes of instruments, flight instruments and engine instruments. Basic flight instruments are the altimeter, turn-and-bank indicator, artificial horizon, airspeed indicator, rate-of-climb indicator, gyrocompass, and magnetic compass. These instruments indicate to the pilot the aspect of the plane and its altitude. The student is trained to fly the plane relative to these instruments, which are more sensitive than direct observation. Examples of engine instruments are the tachometer, fuel pressure gauge, oil pressure gauge, carburetor air intake pressure gauge, manifold pressure gauge, torque meters or output horsepower, and temperature gauges for air intake, oil and fuel, and the cylinder heads. The precise operating status of the engine is determined by adjusting fuel intake and air intake and is indicated by the tachometer, output horsepower, manifold pressure, temperature, etc. From the point of view of overall flight, however, the two critical outputs are the thrust of the propellers and the consumption of fuel.

In normal flight the aircraft is subject to gravity, the thrust of the engine, and the action of the air. One considers the airplane as a solid body consisting of the fuselage, fixed wing surfaces (the empennage), and control surfaces. For the purpose of analysis, the latter are considered to have fixed angular relations relative to the fixed elements at any instant of time, although this relative position is actually controlled by the pilot. According to Newtonian dynamics, the forces acting on the aircraft produce accelerations that can be integrated to yield the motion and position and aspect of the aircraft. Thus, we have a rough tentative diagram for the flight box (Figure 2.3).

We must specify the position and motion of the aircraft by means of certain variables. There are various ways in which the motion of a solid body in space can be described. Our method contains some redundancy but has a compensating flexibility, which is useful. We begin by choosing a coordinate system (the "fixed" system) that is fixed relative to the earth. We will assume that this fixed system is an "inertial frame," that is, a coordinate system in which Newton's law, force = mass × acceleration, is valid; although this is not strictly true because of the rotation of the earth. We also choose a coordinate axis system fixed in the aircraft. It is customary to choose the first, or x-axis, along a longitudinal axis of the aircraft with positive direction forward. We choose the second, or y axis, to be horizontal in normal flight, and the third, or z axis, to be vertical. This coordinate system is called the "body" system.

Let $\mathbf{u} = (u^1, u^2, u^3)$ be the displacement vector of the origin of the body system, expressed by components, in the fixed system. During the flight each component, u^1, u^2, and u^3, will be a function of the time. Let $\mathbf{i}_1 = (i_1^1, i_1^2, i_1^3)$ be the unit vector in the positive direction of the first axis of the body system, expressed by components in the fixed system and let $\mathbf{i}_2 = (i_2^1, i_2^2, i_2^3)$ and $\mathbf{i}_3 = (i_3^1, i_3^2, i_3^3)$ be the corresponding unit vectors along the second and third body axes. During flight, the components i_r^s, r, $s = 1, 2, 3$, are also functions of the time.

The twelve quantities, i_r^s, u^t, are adequate to describe the position of the aircraft: For if A is any point in the aircraft, it will have three constant coordinates (x^1, x^2, x^3) in the body system, and if (s^1, s^2, s^3) is the corresponding set of coordinates in the fixed system, one has

$$(s^1, s^2, s^3) = (u^1, u^2, u^3) + x^1 \mathbf{i}_1 + x^2 \mathbf{i}_2 + x^3 \mathbf{i}_3$$

or $\mathbf{s} = \mathbf{u} + \mathbf{xi}$ when one considers $\mathbf{s}, \mathbf{u}$, and $\mathbf{x}$ as one-rowed matrices and $\mathbf{i}$ is the 3×3 matrix with rows corresponding to the $\mathbf{i}_r$ vectors. We will now drop the

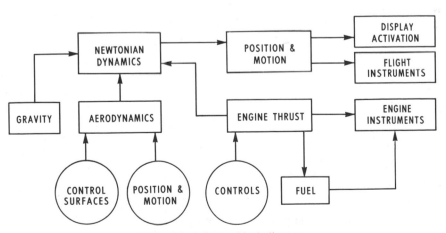

Figure 2.3. Influence block diagram.

use of boldface for vectors and matrices and write $s=u+xi$. Since x is a constant, this equation describes the position of A when the components of u and i are given as functions of the time.

Since x is a constant, we have

$$\frac{ds}{dt}=\frac{du}{dt}+x\frac{di}{dt}.$$

We must express di/dt. Let the prime denote the transpose. Since i is an orthogonal matrix, we have $ii'=1$ and hence

$$0=\frac{di}{dt}i'+i\frac{di'}{dt}=\frac{di}{dt}i'+i\left(\frac{di}{dt}\right)'=\frac{di}{dt}i'+\left(\frac{di}{dt}i'\right)'$$

by the well-known properties of the transpose. Let $\Omega=(di/dt)i'$. This equation can be written $0=\Omega+\Omega'$, or Ω is antisymmetric. Hence, Ω can be written

$$\Omega=\begin{bmatrix} 0 & \omega_{12} & -\omega_{31} \\ -\omega_{12} & 0 & \omega_{23} \\ \omega_{31} & -\omega_{23} & 0 \end{bmatrix}.$$

We also have $di/dt=\Omega i$. If we let ω denote the vector $(\omega^1,\omega^2,\omega^3)$ with $\omega^1=\omega_{23}$, $\omega^2=\omega_{31}$, $\omega^3=\omega_{12}$, then

$$x\Omega=(x^1,x^2,x^3)\begin{bmatrix} 0 & \omega^3 & -\omega^3 \\ -\omega^3 & 0 & \omega^1 \\ \omega^2 & -\omega^1 & 0 \end{bmatrix}$$

$$=(\omega^2x^3-\omega^3x^2,\omega^3x^1-\omega^1x^3,\omega^1x^2-\omega^3x^1)$$

$$=\omega\times x.$$

We have

$$\frac{ds}{dt}=\frac{du}{dt}+x\frac{di}{dt}=\frac{du}{dt}+x(\Omega i)=\frac{du}{dt}+(x\Omega)i=\frac{du}{dt}+(\omega\times x)i,$$

and, for the acceleration,

$$\frac{d^2s}{dt^2}=\frac{d^2u}{dt^2}+\left(\frac{d\omega}{dt}\times x\right)i+[\omega\times(\omega\times x)]i.$$

The velocity of any point can be expressed in terms of du/dt and ω. We can also express the velocity as a vector in the body system by means of v, defined by $du/dt=vi$ and $ds/dt=(v+\omega\times x)i$. Hence the position and motion of the aircraft can be expressed in terms of u, i, v, and ω.

The Newtonian dynamics of a rigid body is usually based on the assumption that the body consists of a large number, n, of particles small enough so that each can be considered located at a point. Let particle α have position vector $s_\alpha = (s_\alpha^1, s_\alpha^2, s_\alpha^3)$ in the fixed (inertial) system, mass m_α, and be subject to a force F_α, which is the resultant of all the forces acting on the particle. Thus, $F_\alpha = m_\alpha d^2 s_\alpha / dt^2$.

If s_0 is the position of the center of gravity, then $\sum_\alpha m_\alpha s_\alpha = M s_0$, where $M = \sum_\alpha m_\alpha$ is the total mass and $\sum_\alpha m_\alpha (s_\alpha - s_0) = 0$. If we sum over all particles, we obtain,

$$F_R = \sum_\alpha F_\alpha = \sum_\alpha m_\alpha \frac{d^2 s_\alpha}{dt^2} = M \frac{d^2 s_0}{dt^2}.$$

In summing the forces, the forces between particles cancel, so that F_R is the resultant of the external forces applied to the set of n particles, that is, to the body. This equation can also be written

$$F_R = M \frac{d}{dt} \left(\frac{ds_0}{dt} \right) = M \frac{d}{dt} (v + \omega \times x_0)i$$

$$= M \left[\frac{dv}{dt} + \frac{d\omega}{dt} \times x_0 + \omega \times v + \omega \times (\omega \times x_0) \right] i,$$

where x_0 is the center-of-gravity displacement vector in the body frame. If G_R is the vector expression for F_R in the body frame, $F_R = G_R i$, and we have

$$G_R = M \left[\frac{dv}{dt} + \frac{d\omega}{dt} \times x_0 + \omega \times v + \omega \times (\omega \times x) \right].$$

We also have

$$F_\alpha - \frac{m_\alpha}{M} F_R = m_\alpha \left(\frac{d^2 s_\alpha}{dt^2} - \frac{d^2 s_0}{dt^2} \right)$$

and

$$\left(F_\alpha - \frac{m_\alpha}{M} F_R \right) \times (s_\alpha - s_0) = m_\alpha \left(\frac{d^2 s_\alpha}{dt^2} - \frac{d^2 s_0}{dt^2} \right) \times (s_\alpha - s_0)$$

$$= m_\alpha \frac{d}{dt} \left[\left(\frac{ds_\alpha}{dt} - \frac{ds_0}{dt} \right) \times (s_\alpha - s_0) \right].$$

We sum over the particles and obtain, since $\sum_\alpha m_\alpha (s_\alpha - s_0) = 0$,

$$\sum_\alpha F_\alpha \times (s_\alpha - s_0) = \frac{d}{dt} \left[\sum_\alpha \left(\frac{ds_\alpha}{dt} - \frac{ds_0}{dt} \right) \times (s_\alpha - s_0) m_\alpha \right].$$

Here again, when summation occurs, the torques due to internal forces cancel. Let G_α be the expression of F_α in the body axis; i.e., $F_\alpha = G_\alpha i$, and recall $s = u + xi$, $ds/dt = du/dt + (\omega \times x)i$. We obtain

$$\sum_\alpha G_\alpha i \times (x_\alpha - x_0)i = \frac{d}{dt} \left\{ \sum_\alpha m_\alpha [\omega \times (x_\alpha - x_0)i] \times (x_\alpha - x_0)i \right\},$$

and since the cross product is invariant under an orthogonal transformation,

$$\frac{d}{dt}\left(\left\{\sum_\alpha m_\alpha[\omega \times (x_\alpha - x_0)] \times (x_\alpha - x_0)\right\}i\right) = \left[\sum_\alpha G_\alpha \times (x_\alpha - x_0)\right]i$$

$$= \left(\sum_\alpha G_\alpha \times x_\alpha - G_R \times x_0\right)i$$

$$= (T^0 - G_R \times x_0)i.$$

Here T^0 is the resultant of the external torques expressed in the body system. The vector

$$P = \sum_\alpha m_\alpha[\omega \times (x_\alpha - x_0)] \times (x_\alpha - x_0)$$

$$= \sum_\alpha m_\alpha(\omega \times x_\alpha) \times x_\alpha - M(\omega \times x_0) \times x_0$$

is termed the angular momentum, and in the limit associated with taking the number of particles indefinitely large, one has

$$P = \iiint \delta(\omega \times x) \times x \, dV - M(\omega \times x_0) \times x_0$$

where $\delta = \delta(x^1, x^2, x^3)$ is the density and the integration is over the volume of the aircraft. P is a vector that depends on the vector ω by a transformation given by a matrix J called the moment of inertia. The formula $(a \times b) \times c = (a \cdot c)b - (b \cdot c)a$ permits one to express J in terms of the matrices J_0 and X_0 and a scalar j_0 defined as follows:

$$J_0 = (j_0^{rs}),$$

where

$$j_0^{rs} = \iiint \delta x^r x^s \, dV, \qquad X_0 = (x_0^r x_0^s)$$

and

$$j_0 = \iiint \delta[(x^1)^2 + (x^2)^2 + (x^3)^2] \, dv - M[(x_0^1)^2 + (x_0^2)^2 + (x_0^3)^2].$$

Then $J = J_0 - MX_0 - j_0^1$ and $P = \omega J$. Our previous result becomes

$$\frac{d}{dt}(\omega J i) = (T^0 - G_R \times x_0)k$$

or

$$\left[\frac{d\omega}{dt} J + \omega \times (\omega J)\right]i = (T^0 - G_R \times x_0)i.$$

This yields a relation to specify $d\omega/dt$.

Thus Newtonian dynamics yields a set of equations to specify $d\omega/dt$, dv/dt, du/dt, and di/dt, i.e.,

$$\frac{d\omega}{dt} J + \omega \times (\omega J) = T^0 - G_R \times x_0$$

$$G_R = M \left[\frac{dv}{dt} + \frac{d\omega}{dt} \times x_0 + \omega \times v + \omega \times (\omega \times x_0) \right]$$

$$\frac{du}{dt} = vi, \qquad \frac{di}{dt} = \Omega i$$

if one knows the values of T^0, G_R, x_0, and of course, ω, v, u, and i. This is then a set of differential equations that gives the time rate of change of the eighteen components of w, v, u, and i in terms of themselves.

There are numerical procedures, which, when the value of w, v, u, and i are given at a time t_0, yield the values of these same quantities at times $t_0 + h$, $t_0 + 2h$, $t_0 + 3h$, ... with high accuracy if a sufficiently small value of h is used. This is termed step-by-step integration with step h.

We can now complete our discussion of two of the boxes in the block diagram of Figure 2.3 by stating the associated variables and how they are to be determined mathematically as functions of the time. The variables for the Newtonian dynamics are the derivatives $d\omega/dt$, dv/dt, du/dt, di/dt, and we have equations of these in terms of T^0, G_R, and x_0, which presumably will come from other boxes. The variables for the position and motion box are u, i, v, and ω, and these are to be obtained by numerical integration of the differential equation system.

There are three major types of forces acting on the body. There are three contributions to G_R. One of these is gravity, which in the fixed system is a vector straight down. However, in the body system it is a vector g, which satisfies the equation $gi = M(0, 0, g)$. The contribution of gravity to $T^0 - G_R \times x_0$ cancels out and can be omitted from T^0 if it is omitted from $G_R \times x_0$.

The thrust due to the engine is proportional to the power output, and the computations needed to specify the latter in time are quite complex. But we can, for our purposes, assume that the power output is a function of the altitude, the fuel-feed setting, the air-intake setting, and the speed of the aircraft. We can consider the thrust as a vector in the form

$$G = C(u^3, v, f_t, f_a)(1, 0, 0);$$

i.e., the propeller axis is parallel to the x axis of the body system. If this axis does not pass through the center of gravity, there will also be a torque around the y axis due to the engine thrust, T^e. The fuel consumption is also an output from the engine box, and the position of the center of gravity is determined by the amount of fuel available.

It remains to consider the aerodynamics. If we consider the atmosphere as still and the airplane as moving through it with velocity v, the aerodynamic forces presumably are the same as in the case in which the aircraft is in a fixed position and subject to a wind opposite in direction to v. Thus, the aerodynamics are determined by the vector v. In normal flight, v is close in direction to the positive first axis of the body system, and its position can be specified as follows. Take a vector of the appropriate length along the positive first axis. Rotate this vector by an amount α around the second axis in such a way that a small positive α yields a negative third-axis component. This α rotation can also be considered as a rotation of the body-system axes. One follows this by a rotation of an amount β about the new position of the third axis. This specifies the position of v in terms of the angles α and β, which are small in normal flight.

The aerodynamic forces and torques are determined empirically in a wind tunnel. In this device, the "wind" is fixed in direction and the instrumentation is also fixed. A model of the aircraft with the first body axis directed positively into the wind and the second body axis horizontal determines a system of coordinates called the "wind system." Measurements are performed corresponding to various values of α and β by rotating the model inversely to the above procedure for locating the vector v. However, the measurements are still in the wind axis system and to yield vectors in the body axis system one must perform a rotation with matrix $h = h(\alpha, \beta)$. Thus if H is a vector in the wind axis system and G the corresponding vector in the body axis system, one has $Hh = G$.

The measurements yield approximate expressions for the aerodynamic forces and torques in a number of variables. One needed quantity is the density of air, ρ. Within the practical limitations of the current problem, ρ can be considered to be a function of the altitude, u^3. The density of air yields the speed of sound and Mach number, Ma, which is the ratio of $|v|$ to the speed of sound. The aerodynamic forces are also dependent on the values of the angles of the control surfaces, δ_R, δ_B, δ_E, δ_A, for rudder, dive brakes, elevators, and ailerons, respectively, and on the slot position variable δ_F. Let S denote the wing area, b the wing space, and c the mean aerodynamic chord.

The three components of the aerodynamic force, H_a^1, H_a^2, H_a^3 are represented by formulas of the following type:

$$H_a^1 = \tfrac{1}{2}\rho v^2 S[C_x(\alpha, \text{Ma}) + C_x(\beta) + C_{xF}\delta_F + C_{xB}\delta_B]$$

$$H_a^2 = \tfrac{1}{2}\rho v^2 S(C_{y\beta}\beta + C_{yR}\delta_R)$$

$$H_a^3 = \tfrac{1}{2}\rho v^2 S[C_z(\alpha, \text{Ma}) + C_{zE}\delta_E + C_{zF}\delta_F].$$

The functions $C_z(\alpha, \text{Ma})$ and $C_x(\alpha, \text{Ma})$ are the empirically determined lift and drag functions, respectively. For Ma fixed and relatively small values of α, $C_z(\alpha, \text{Ma})$ is an increasing function of α but reaches a maximum and abruptly drops, corresponding to the condition of stall.

The aerodynamics torques are quite similar:

$$S_a^1 = \tfrac{1}{2}\rho v^2 b S (C_{lA}\delta_A + C_{l\beta}\beta + C_{lR}\delta_R) + \tfrac{1}{4}\rho v S b^2 (C_{lP}|P| + C_{lw^3}w^3)$$
$$S_a^2 = \tfrac{1}{2}\rho v^2 c S [C_m(\alpha, \text{Ma}) + C_{mE}\delta_E + C_{mF}\delta_F] + \tfrac{1}{4}\rho v c^2 S (C_{m\dot{\alpha}}\dot{\alpha} + C_{mw^2}w^2)$$
$$S_a^3 = \tfrac{1}{2}\rho v^2 b S (C_{n\beta}\beta + C_{nR}\delta_R + C_{nA}\delta_A) + \tfrac{1}{4}\rho v b^2 S (C_{nw^3}w^3 + C_{nP}|P|).$$

From these, the corresponding quantities in the body system are obtained by means of the matrix h; i.e., $H_a h = G_a$, $S_a h = T_a$.

Thus, the resultant force in the body system is given by $G_R = g + G_e + G_a$, and if g is omitted in $G_R \times x_0$, one can use $T_R = T_a + T_e$. The lift term corresponding to $C_z(\alpha, \text{Ma})$ must balance g essentially. However, associated with the lift, there is a drag term $C_x(\alpha, \text{Ma})$ and a moment around the y axis. The engine thrust must balance the drag. To counteract the moment around the wing, the stabilizer or horizontal tail surface produces a counter torque. The resultant of these two torques is represented by the $C_m(\alpha, \text{Ma})$ term in the S_a^2 expression. Thus, if lift equals gravity, engine thrust equals drag, and the turning moments cancel, one can have straight horizontal flight at constant speed with the control surfaces in neutral position. A rotation around the first body axis is called roll. The ailerons are used to counteract unwanted roll and to permit roll adjustments. Turns in a horizontal plane involve the rudder to a certain extent, but mostly a roll adjustment so that the lift has a horizontal component in the direction of turn. To ascend, extra lift in the H_a^3 component is provided by the elevators and flaps or slots. These also produce torques around the second body axis (y axis). A rotation around the y axis is called pitch and one around the z axis is called yaw.

The purpose of the various coordinate systems is now clear. The inertial system is required to establish the position of the plane and to apply Newton's laws. In the body system, the moment-of-inertia matrix is constant, and integration is performed in this system. The aerodynamics is given in the wind system.

We are now able to complete the normal flight block diagram and, by implication, the math model (Figure 2.4). This is to be considered as part of Figure 2.2. In this block the motion and position develop according to the system of differential equations given above. We have other systems of differential equations for the phases in the usual flight sequence. For example, the first takeoff phase in which all the wheels are on the ground and the second takeoff phase in which nose wheel is off the ground. One can also have a disaster if the wing scrapes the ground in takeoff. This must also be considered a flight phase and simulated.

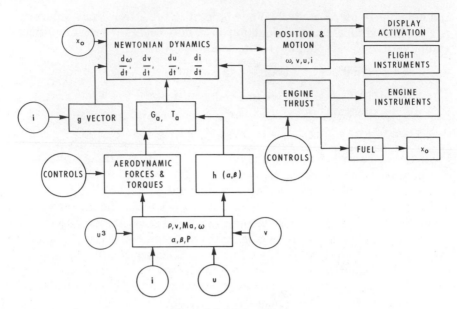

Figure 2.4. Normal flight.

Thus, the flight of the aircraft can be described by various regimes of development, governed in most cases by systems of differential equations. A change of regimes corresponds to a change of state variables. Such a change will occur when a variable or combination of variables satisfies a relation or inequality. For example, normal flight can end in a stall if α becomes too large, or takeoff two becomes normal flight when u^3 reaches a certain value. Thus, each regime proceeds until such a criterion is satisfied.

The controls in an aircraft cockpit operate power systems that position the control surfaces. These power systems function against aerodynamic forces depending on the plane's velocity, altitude, angular motion, and trim. Thus, the angular positions of the control surfaces are governed by differential equations, which, in general, must be integrated numerically.

The display system is concerned with landing and needs only to reproduce effective cues such as the edges of the landing field. One may wish, however, to add background details. In general, the information to be displayed corresponds to points or lines fixed in the inertial coordinate system. One possibility for the display is to assume that it corresponds to a plane fixed relative to the body axis system, perpendicular to the x, or first, axis and, appropriately, in front of the aircraft. For each object or point to be displayed, one must determine where the line of sight from the pilot to the object or point intersects the display plane. This is an interesting mathematical relation to work out. One would normally expect all such computations to be executed in a display computer that receives u and i from the flight computer.

2.7. Block Diagrams

Originally, the term "block diagram" referred to the description of the relation between the various pieces of equipment of a complete device. Each component is indicated by a box, and the interconnecting lines represent either a mechanical connection such as a rotating shaft or an electrical connection. When a simulation is embodied in a complete device, as, for example, in a flight trainer, the term is still used in this way. If one of the ultimate aims of an engineering project is the development of a physical system or device, an essential output is a block diagram in this sense. We will refer to this as an equipment block diagram.

The concept of an "influence block diagram" for analyzing the situation to be simulated arose in the use of analog equipment for simulation purposes. In an analog simulator each mathematical operation is represented by a special component, and each formula or computational procedure is represented by equipment in the same rack or associated racks. Thus, if the flight of an airplane is to be simulated, one would have groups of racks corresponding to the computation of the aerodynamic forces on the plane or the action of the engine or of the kinetic integrations and changes of coordinate systems. For a discussion of analog computation, see Murray.[5]

Correspondingly, if one is planning to set up such a system, one would analyze the original situation into elements that could be represented by coherent blocks of computation in the analog sense. The relationship of these elements can then be described by a block diagram that is analogous to an equipment block diagram. The use of block diagrams in this way is an effective method for carrying over past experiences with similar problems and incorporating improvements in understanding.

When digital data processing is used instead of analog processing, blocks of coding in the machine program replace combinations of analog equipment. In programming, it is desirable to retain this resolution of the program in coherent blocks, i.e., the "modularity." Thus, one obtains program block diagrams, corresponding to the influence block diagram of the original situation. This digital programming block diagram has connections corresponding to transfers of data, and this is similar to the analog case. It is, of course, different from the flow chart in digital programming, which indicates sequential relations in the execution of the program in time. The greater flexibility of digital programming permits wider-ranging simulations so that the possibilities for influence block diagrams are greatly expanded and more complex. However, the logical possibilities for the influence block diagram must be limited to the requirements by practical considerations, and it is desirable to remain as close as possible to previous experience.

We have indicated the role of tradition in setting up the influence block diagram. However, the simulation must also satisfy certain objectives, and this shows the importance of a precise requirements document. A simulation should not be more complex than that required by the objectives. This is generally reinforced by limitations on available resources for the simulation, such as manpower, computation, and total project time (i.e., deadlines).

For these reasons, therefore, the influence block diagram will represent compromises in the amount of detail represented and the completeness of the quantitative description of the elements. It is in general illuminating to discover whether the local simulation tradition had an analog epoch. We notice two other uses of the term "block diagram," i.e., equipment block diagram and program block diagram.

The format of block diagram, math model, flow chart, and scenarios appears quite abstract. But actually, this represents an engineering tradition that originated in modeling in the literal sense. Patent procedures at one time were dependent on "working models," and models are still used for ships, public works, and to a certain extent aircraft. The differential analyzer permitted one to replace an actual model by a configuration of computing components that solved the differential equations satisfied by the original. The math model for this situation was, of course, the system of differential equations.

The idea of a mechanical differential analyzer was proposed by W. Thomson (Lord Kelvin)[6],[7] in 1875, but there were technical difficulties that were overcome only in the late 1920s. These devices came into use in the decade before World War II (see Bush[1]). Electrical components for analog computation were developed in World War II by Bell Laboratories. Electrical differential analyzers were much easier to program, and abrupt changes such as those corresponding to a change of flight phase could be introduced by relays.

The ability to solve differential equations permitted one to replace models for engineering purposes by models consisting of computing components, which were more readily constructed. Furthermore, variations in design and environment could be tested in an inexpensive fashion. The use of analog differential analyzers continued over a period of possibly thirty years and resulted in a general acceptance of analogy based on mathematical equivalence. By the middle sixties, digital data processing was capable of competing with analog equipment on a cost-effectiveness basis and the logical flexibility of digital equipment permitted a considerable expansion of the simulation field.

The use of diagrams to describe relatively complex situations is not confined to the examples quoted above. An electric circuit diagram is essentially an "influence block diagram" with special rules for determining the math model. Complex mechanical devices can be analyzed by the equivalent of a circuit diagram as well as electromechanical systems. One very striking type of diagram is that which is used to indicate the sequence of chemical reactions that are associated with certain activities of living cells. This type of diagram has blocks for chemical reactions and connections for the chemicals themselves. Similar diagrams are applicable to industrial processes. With the current interest in ecology, diagrams associated with the biological environment are often used, and diagrams are obviously useful in economics and sociology. In general, in a diagram the blocks should have the same significance as should the connections.

2.8. Equipment

An engineering simulation must be developed in phases that may overlap but that will permit the orderly use of resources. The first phase must involve the analysis of the original situation and yield the math model that governs the time history of the vicarious experience. Another phase must deal with equipment—both the data processing system and that which interacts with the subject. A third phase is programming, and a fourth phase consists in integrating the program and the diverse equipment into an effective device. The fifth phase is the operational one that will lead to the appropriate conclusions.

In practice, one can divide the equipment into three systems, i.e., data processing; an input system, which produces data for the data processing; and an output system, which is controlled by data processing. In certain cases all three systems may be large. We will refer to the data processing system as the "computer" but, in addition to the central processing unit and storage hierarchy, it will usually contain peripheral equipment such as hard-copy printers, card readers, and card punches.

The high speed of electronic circuits favors the dynamic type of simulation in which a complete cycle of computation is repeated many times a second. Many of the numbers are intermediate, and a smaller number uniquely determines the time history. Similarly electronic displays such as television pictures or cathode-ray-tube outputs generate complete pictures or "frames" thirty times a second. Most of this information in the display does not vary from frame to frame. Such displays normally have their own store of information associated with them and with it a specialized computer to incorporate the new information as required.

Thus, the calculation and the display storage are two disjoint banks of information with a special problem of intercommunication. The communication process is itself demanding in logical capability, and the computer is used for both computing and communication. The two processes are interlaced in time.

In addition to the above displays, mechanical motion may also be desired as output. An electrical signal from the computer is used to produce this motion either directly, as in meters, or through amplification to attain the required power. When such an electrical signal has a continuous range (e.g., a voltage) it is called an "analog" signal. Normally a special "interface" with the computer is needed to produce the set of analog outputs required during each cycle of computation. In the interface the continuous signal is produced from a digital value obtained in the calculation. Audio effects may also require analog signals for control.

The requirements on the input system are similar. Thus, the various controls on a flight trainer have continuous ranges, and the value in each case must be expressed in binary digital form and introduced as such into the computer. A simulation may require an enormous amount of input information in total, but only a small fraction of it at any specific time. A simulation may require extensive map information for its entirety but only local map information at any instant of time. The amount of information may require that it be stored outside the direct-access memory of the computer. Under these circumstances one needs an additional computer to select and transmit the immediately needed information. This additional computer receives guidance information from the main computer.

The second phase, then, consists of decision on these three systems in order to attain the objectives of the simulation. The objectives determine the output displays and analog input and output requirements. The precise choice of input and output equipment must also take into account availability, reliability, and company policy. The decision on input and output equipment determines precisely the input data available for the computation and the output data required from it. This information and the math model specify the amount of computing capability required for the data processing.

The choice of the equipment can be represented by an equipment block diagram in which the connections indicate a transfer of data and can be associated with data communication rate. If in addition we associate with the block for each piece of equipment a data processing capability, we can consider that the equipment block diagram is also a data flow diagram. This constitutes a diagrammatic summary of the second phase.

2.9. The Time Pattern of the Simulation

The third stage in the development of the simulation is programming. This is an engineering effort of some magnitude and must be carefully phased. A possible phasing would be (a) requirements, (b) structuring and specifications, (c) coding, and (d) testing.

The requirements of the program have two interacting aspects, one generated by the math model, the other by the data flow needs of the equipment. These aspects are reconciled by determining (a) the time pattern for the computation, (b) accuracy and resolution specifications for the computing procedures to realize the math model, and (c) testing exercises, including those to be used in assembling the equipment into a complete device.

The time pattern for the computation must reflect the time development of the original situation, provide for the reception of input, and produce the activating output. Thus, one must have regimes of continuous change terminated by critical events in which there are abrupt changes of state variables.

There are two ways in which this pattern is represented in the computer. If the development of the continuous parameters can be directly extrapolated, one can have a critical event simulation. In this type one has a succession of critical events, and at each such critical event the time of the predictable future ones are computed. On the other hand, if the change in the continuous variables is determined by a system of differential equations, the system must be solved by step-by-step methods, which advance the time by small fixed time increments. After each such advance one can test whether a change of the state variables will occur in the next time step and adjust the advance up to the critical point. In most cases a somewhat simpler procedure is acceptable in which one tests whether a critical event has occurred in the last interval, and if so, one assumes that the time of the critical event was at the end of the interval.

In general, the analysis of the situation may be difficult to the point that given the present aspect, one can make predictions only for a short interval into the future. However, the fixed-time-increment procedure and automatic data processing, which can be programmed to iterate such predictions, yields numerical simulations under these circumstances. The other extreme would be a case where the variation of the continuous variables corresponding to any desired length of time and at each critical event could be computed, the time of occurrence of the next critical event could therefore be computed directly. For example, in a maintenance simulation, the time to next failure is usually determined by a probabilistic method.

The programming pattern for a critical event simulation is based on a list of anticipated critical events arranged by time of occurrence. If the simulation is not in real time, time is advanced by taking the next critical event in time and computing those future events that are predictable at this point in time and adding to the basic list. In a real-time simulation, a computer clock is tested at intervals until the next critical event in time is scheduled to occur. The more general situation requires the "fixed time interval approach." The continuous variables are divided into sets; one set is determined by a system of differential equations, another set may advance kinetically, and a third set may consist of functions of variables in the first two sets. Thus, at each time increment the advanced values of each set are computed successively and then the possibility of a critical event is determined. When a critical event occurs, usually some computation is required to determine the new values of the state variables and to make the adjustments needed to initiate the new regime of time-increment advance.

Input and output also determine characteristics of the time pattern of the computation. Consider the fixed time interval case. Input normally corresponds to the values of certain parameters, and these should remain constant during the computations associated with a given time value. Inconsistencies in the values of the parameters may well have objectionable mathematical and systemic effects. Thus, the input values must be buffered. When they are brought into the computer, they are placed in certain registers and introduced into the computation at the beginning of the computation for a fixed time value. These inputs can be introduced into the computer by means of an "interrupt" capability, which will transfer the input to the buffer registrars at any time without affecting the computation. Otherwise input periods are interspersed through the computation.

It is a characteristic of output that data is required on a fixed schedule. Consequently, the computation for each fixed time increment may be segmented so that between each segment a certain amount of output is processed and communicated.

The requirements for advancing the variables and input and output indicate the time pattern of the computation for a fixed time interval simulation. Relative to a critical event simulation, the possibilities are more complex. Input may only be required at each critical event, and if the simulation corresponds to real time, then the procedure at each critical event begins with an input sampling. It is also possible that certain inputs trigger critical events in addition to the precomputed events on the list. Certain outputs may be associated with critical events and occur with the corresponding computations. Scheduled output may be handled by an auxiliary program independent of the main simulation program. This may correspond to a special capacity of the computer to handle two independent programs, or it may be a specialized programming arrangement based on a computer clock.

Thus, the time pattern of the data processing is determined by the time pattern of the original situation and the input and output requirements. This time pattern may be realized by a sequence of instructions in the computer, i.e., by using the normal logical capability of the computer with perhaps a clock feature. However, they may also utilize special interrupt or multiple-programming features. The kinetic or dynamic character of the math model determines the major possibilities for the programming, but this must be supplemented in order to provide input and output. In the computation a continuous regime consists of a sequence of central processing instructions dealing with a change of the continuous variables. The end of such a computation regime corresponds either to a critical event in the previous sense or to input or output. Clearly we can represent this computation pattern by a flow chart. The special interrupt or multiple-programming features relieve the need to make interruptions in the sequence of computer instructions and tend to make the program flow chart similar to the situation flow diagram.

The output devices require not only data rates but also accuracy and resolution. The latter in turn are a requirement on the computation procedures that realize the math model. Like the computation flow chart these also must be documented.

The program flow chart and the possible time patterns in the computation are usually so complex that it is not practical to test all the logical possibilities. Consequently a certain number of patterns must be chosen to yield overall systems tests that will be reasonably conclusive.

2.10. Programming

It is of great practical importance to have the computer program in "modular form." A module may correspond to a continuous program regime or to subroutines, which may appear in a number of such regimes. Other modules may correspond to input or output processing or the computations associated with critical events. The program flow chart can be realized by appending to each module a decision block that will determine the next module. But greater flexibility is obtained by the use of an "executive program" to which the computation control returns after each module has been executed and where the next module to be executed is determined. The "flow chart" in the previous sense is now represented in a compact form in the executive. In this form, it is practical to use much more sophisticated flow charts and adjust them on the basis of experience. Each module must be computationally complete in itself and one cannot have numerical subresults dangling from one module to the next to avoid duplication. This juggling of numerical subresults was a darling of flow chart programming and made readjusting the program a hazardous process for anyone including the original programmer.

To illustrate these ideas let us consider in particular the programming of the OFT discussed in Section 2.5. In the usual flight phases the time development of the simulation is given by a system of differential equations. A numerical procedure is used to obtain the solution at times $t_0, t_0 + h, t_0 + 2h, \ldots$, and the "step size" h is also the time increment by which the simulation advances. (A multiple of h can be used as time increment to advance the simulation.) In order to obtain the required accuracy, this time increment must be chosen small, for example, 0.05 seconds.

There is also considerable input and output that must be processed in each time increment. The input and output apparatus external to the computer is slower than the internal procedures of the computer, so that one cannot process input and output in one batch at each time increment. Thus, while the simulation is being advanced for a specified time increment, the input for the next time increment is being placed in an appropriate buffer in batches. Similarly the output from the previous time increment is being processed and sent out of the computer in batches. Because of the brief increment time, this phasing of input and output is not noticeable to the trainee.

The total program appears as a complex of modules and the executive program that controls the sequence of execution. One can readily classify the types of modules needed, and this classification essentially determines the structure of the executive program. One has:

(a) Input modules that operate on the information in the input buffers and introduce it into the program in an appropriate format and scale.

(b) Output modules that express numerical information in the required output format and buffer it for the output process.

(c) Testing modules for each flight phase that determine whether a change is to occur and the new flight phase.

(d) Initiating modules that set up the computing process for a newly initiated flight phase. A new flight phase may use quantities not used in the previous phase, for example, and initial values for these may have to be computed.

(e) Computing modules.

The modules usually have to be executed in a definite sequence. Thus, the air density, Mach number, velocity, α, β, etc. must be computed before the aerodynamic forces and torques in the wind axis system. The latter in turn are used to compute the corresponding quantities in the body system. When a numerical integration procedure is applied to a system of differential equations, all the derivatives must be computed at the same value of the independent variable, i.e., the time, in order to validate the customary error analysis and the choice of certain constants in the integration formulas.

Thus, in a given time increment, the first choice of the executive program will be a type (a) input program. This will be followed by a type (c) testing program to see if a change of phase is required. If a change occurs a type (d) initiating program will follow. After that one will have a sequence of type (e) computing modules interspersed with input and output modules. The sequence is terminated by an output processing, type (b), module.

The resolution of the program into modules and the formulation of the executive program constitutes a structure for the computation that must be precisely documented. The logical requirements of the executive program must be specified as must the accuracy, resolution characteristics, and data rates for the calculations in each module. The overall test exercises previously determined imply specific test exercises for the executive program, and it is usually desirable to expand these so that one obtains a sequence of tests that can be applied in a progressive manner, starting with the most straightforward substructures. Similarly a system of tests for each module should be set up, structured on the tests implied by the overall system testing. These aspects should be incorporated into an overall structure document and specification documents for the executive program and the modules.

Usually it is not possible to solve the mathematical relations in the math model by a finite arithmetic algorithm. Thus, one must use approximating numerical procedures in the computation. These procedures must have the required accuracy, must be stable, and must be applicable over the range needed in the simulation. The accuracy is specified by the data requirements and this also applies to the range. Stability refers to the reliability of the numerical procedure. Many of these procedures have a range of starting situations, but only a portion of this range will produce a reasonably accurate answer. The reader is familiar with the Newton–Raphson method for solving equations. Some methods for solving differential equations are always unstable, but others are stable if one limits the amount by which the independent variable is extrapolated at each step. The last limit depends on the differential equations to be solved.

Numerical analysis provides many methods for obtaining approximate computational procedures. However, the associated discussion is usually not adequate to determine whether the requirements of the given situation are satisfied. Instead one must make tentative choices and then test the result for such properties as accuracy, stability, and range. These test can be run on a general-purpose computer, not necessarily the computer to be used in the simulation. This experimental programming will also yield the amount of computation needed to attain the desired properties. In general, the more accuracy that is needed, the more computation is required.

Numerical analysis is an essential guide for these experimental computations, and experimentation without this guidance is wasteful of time and yields nondependable results. On the other hand, a preliminary mathematical analysis can be objectionable, yielding far too conservative estimates for accuracy and stability, and is inserted in the development so as to increase project time. The term "programming" may include the documentation of the block diagram, math model, etc. of the original situation, and it usually includes the process of determining the mathematical procedures to be used in the program. The description of these mathematical procedures is referred to as the math model for the computer program.

The experimentation on which the choice of the computer math model is determined utilizes computer languages such as PL1 or Fortran. This experimentation may be continued into a first version of the major aspects of the simulation program. This first version can usually be debugged in a rather straightforward way using the general language facilities. The machine language program produced by a compiler is relatively inefficient, but techniques exist for improving the efficiency of such a program. Alternatively, after the precise procedures are determined, the program can be written directly in an assembler language. The final efficient machine language program can be debugged by means of numerical values obtained from the compiler program. It is desirable to obtain a precisely documented program math model and to base the actual coding on it.

The equipment is usually obtained in the form of a number of components that are assembled in a progressive manner into a complete device. Programming is used to test this integration process as it proceeds step by step.

2.11. Management Considerations

Figure 2.5 is a block diagram that indicates the logical sequence in the development of a simulation. The logical sequence is also the planned sequence of management on the assumption that the amount of experience available will be adequate to limit readjustment and recycling. The little circles are critical events and are usually associated with reporting requirements, although the reporting requirements are more extensive than those indicated.

Planning requires a report schedule and estimates of the various resources needed. Figure 2.6 is a report schedule. Connections could also be drawn to indicate the dependence of later reports on a given report so that a separate table could be set up. The purpose of these report connections is to indicate the effect of a failure to maintain schedule. They indicate that the work on different aspects is subdivided and tightly interconnected to minimize the total project time.

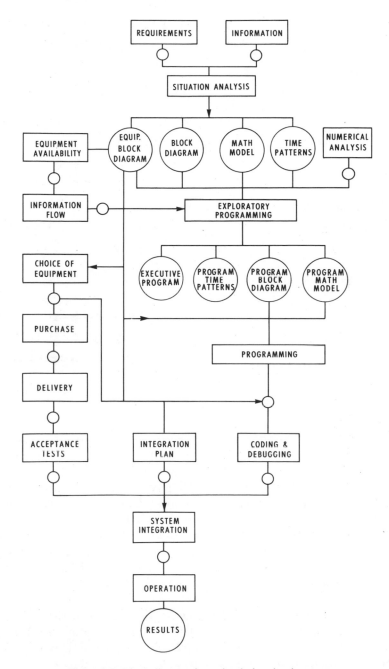

Figure 2.5. Block diagram for a simulation development.

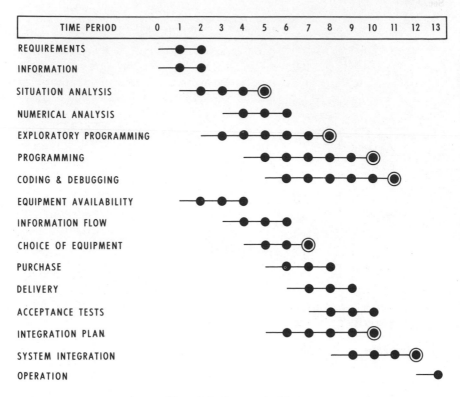

Figure 2.6. Report schedule.

A block-by-block scan of Figure 2.5 and the timing indications of Figure 2.6 can be used to estimate the resources required.

Senior personnel is usually assigned individually on a month-by-month basis, and the assignment of associates follows. This applies to the categories of supervisors, analysts, and engineers. In addition, month-by-month estimates are made of the number of various types of support personnel, such as coding assistants, key punch operators, technical personnel, and secretaries. One must also schedule computer time, technical facilities, office space, and material. This planning permits cost estimates on a scheduled basis.

Some understanding of management procedure is desirable for the professional employee, but our immediate purpose is to indicate the critical requirements for scheduling and documentation. Failure to maintain schedule not only leads to cost overruns, but manpower availability binds and similar binds for facilities and equipment. Diagrams similar to Figure 2.6

with the appropriate dependence relations are used to monitor the time development of the project.

Notice that Figure 2.5 is a block diagram in the influence diagram sense and also in the sense of logical dependence. With appropriate connections, Figure 2.6 is a temporal flow diagram. The overlap in the time diagram, which seems inconsistent with logical dependence, is permissible because a preliminary report on one block may permit work to be initiated on a logically consequent block. This overlap may be highly desirable both for total time and the availability of certain resources. Overlap may also be helpful in both anticipated and unanticipated feedbacks in the total development.

2.12. Validity

A simulation is a contrived experience reflecting some original situation. We have seen the procedures used to set up the simulated situation, and it is clear that the faithfulness of the result depends on the scientific and technical understanding of the original situation.

In general, the scientific principles that govern the situations of interest are well known. But the complexity is in most cases so great that a complete scientific understanding in the sense of a mathematical description is impossible. The more limited mathematical formulations that are available usually cannot be solved. However, the development of large-scale computing facilities made it possible to deal with a considerable range of technical understanding in the form of a combination of scientific theory and empirical information. The empirical information contains direct quantitative data, but it also contains mathematical relations and behavior patterns, for example, stress and strain relations in materials. It is this combination of scientific theory, empirical experience patterns, and empirical quantitative information that is integrated into the math model.

Thus, over a considerable practical range technical understanding cuts through the Gordian knot represented by our lack of complete scientific analysis and mathematical solutions. But one can also consider the situation from another point of view. The term "technical understanding" can also be considered to include experience patterns that do not have a mathematical formulation in the usual sense. Many human activities were based on this technical understanding, including agriculture, the production of metals, and the construction of carriages and sailing vessels. One can consider the introduction of scientific theories as an extension and structuring of previous technical experience, permitting a considerable increase in accomplishment. Metallurgy was tremendously improved by the knowledge of the microscopic structure of metals and the phase characteristics of alloys. Modern theories of heredity greatly assisted agriculture and husbandry. The Wright brothers invented the airplane using Newton's law and empirical information about the action of airfoils and the Otto engine. Thus, one can consider the scientific element in technical understanding as a growth factor that enhances the available procedures or permits a significant breakthrough, not a weakened version of a perfected understanding. The focus of consideration should be the understanding of the complete activity, not a theoretical scientifically complete background. Actually this is a justification for increasing the scientific understanding. This increase takes on many forms, from the purely academic to what may be termed "technical research." Since the objective is the math model for prediction and control, technical research deals with an understanding of experience patterns combining scientific theory and empirical behavior. For example, technical research deals with fluid flows and the behavior of substances under heavy stress empirically, but in a framework of dimension theory and thermodynamics.

This nature of the technical understanding is reflected in many aspects of these efforts. For example, organized team experience becomes extremely valuable. It permits one to begin with a general structure of understanding based on past experience. Of course, the other side of the coin is the capability of adding new aspects to this understanding. One very obvious element in modern technology is the intrusion of additional scientific theories. The team understanding must be responsive in this direction also. But technical understanding also contains an *ad hoc* element in its choice of experience patterns to determine the math model. This implies that there is an essential requirement to test the understanding against experience. One can begin, we hope, with a good approximation to the desired math model, but experience will practically always show a need for adjustment. Experience with a prototype invariably results in design changes and changes in the math model. A highly significant aspect of the use of digital computers in various devices is that experience patterns with these computers have become stabilized and that they do not require "engineering."

This need for readjustment and the ever present possibility that a project may be shown to be not feasible and must be terminated has serious practical consequences. People with poor technical understanding invariably consider a venture that has to be terminated as a "waste." This type of judgment is analogous to overconservative bridge-game bidding. It permits no innovative explorations to take advantage of new developments. No bet is ever made because the stake may be lost.

Technical understanding is the basis for the math model and the faithfulness of the simulation. However, one must also take into account the fact that faithfulness of the simulation is not the primary objective. These simulations are part of a larger project, and the main purpose is to permit the effort to proceed. This continuance requires that certain decisions be made and justified. However, this justification may not need, logically, that a faithful representation be obtained, or one may be willing to proceed on a basis that is not logically complete. The simulation cost must be weighed against the risk of a limited answer.

We have noted at least three types of understanding that will permit such an effort to continue. But even the most cursory technical decision will be based on some quantitative estimates, and thus these understandings are mathematically formulated. Quantitative understanding yields prediction and control and is an essential part of large efforts. It is the most characteristic aspect of our culture, and its complexity reflects forty centuries of development. The applied mathematician will find that there is no simplistic philosophical approach that is a satisfactory substitute for an appreciation of the historical development.

Exercises

2.1. Consider the following devices from the point of view of simulation in terms of (i) different regimes of activity, (ii) snapshot block diagrams for an instant in time in a regime, (iii) critical events, (iv) overall block diagram, (v) flow chart, (vi) math model:

    (a) The mechanism for controlling the water action in a water closet
    (b) A spring-driven or weight-driven clock or watch
    (c) The mechanism of an automatic-fire weapon such as a repeating rifle or machine gun
    (d) The internal ballistics of discharging a gun
    (e) A one-cylinder, two-sided piston steam engine
    (f) An ac–dc electric motor
    (g) Ac only electric motors
    (h) A two-cycle one-cylinder gasoline engine
    (i) A four-cycle one-cylinder gasoline engine
    (j) A rotary internal combustion engine
    (k) A rotary printing calculator that adds, subtracts, totals, subtotals, and clears
    (l) An electronic flip-flop circuit in multivibrator mode; also in one-shot mode

2.2. Analyze the procedural steps required in planning the purchase of a new home, a new car, or a major appliance. What is the relation of advertising to an analytic approach to purchasing?

2.3. The research, development, and design phases of a project produce nothing tangible, so one can cut costs by minimizing these. Discuss the proposition that the least expensive project will be the one with the cheapest research and development stages, and support with reference to actual cases.

2.4. Cost estimates for the total project are given in the requirements stage, research stage, prototype testing, and all through the production phase. What happens to these estimates and why?

2.5. Compare theoretical and empirical methods for investigating steady fluid flows and shock waves.

2.6. For debugging purposes it is desirable to specify the scenarios that cannot occur or are impermissible. One can assume "start" and "end" as critical events, connected to certain other critical events by regimes called "initiation" and "termination," respectively. Permissible scenarios must connect "start" to "end" and can be subject to other restrictions. One question of interest is to replace a given flow chart and permissibility limitations by a flow chart with the same labels for critical events and regimes in which all scenarios correspond to permissible scenarios of the original problem. What axioms for defining a concept of the set of permissible scenarios will yield this result?

2.7. Discuss the scenarios for the devices in Exercise 1. Express the desired actions in terms of scenarios and also the unsatisfactory actions. What options does one have to insure that only the desired action will occur?

2.8. Consider a business situation in which four types of events occur essentially by chance and each type produces a characteristic piece of information, i.e., a data record, that must be processed. Let p_i be the probability that a given event be of type i and a_i the corresponding processing time in milliminutes. The number of events per minute $N(t)$ can be considered as determinate as a function of time of day and season. The data processing system contains an interrupt feature that, independently of the normal processing, places each data record in an appropriate file and maintains files of arrival times and number of records to be processed. The processing may be subject to various requirements, such as a priority system or that of minimizing the maximum waiting time. Discuss an executive-type program for handling this situation and modifications of the notion of scenario to handle situations involving chance.

2.9. In a "map war game," a conflict between two opposing forces, (traditionally, blue and red), is represented as follows. Each side has a command team consisting of the commanding officer and his staff. Military units are represented on a map. The effective map, however, is viewed only by a third group, "the umpires," who also have manpower and logistic records for each side. Each side communicates with the umpires in a form corresponding to the issuance of orders to its units and the return of intelligence information from its units. The umpire group continuously decides what happens in regard to motion, firepower, casualties, and logistics as a result of the orders and the map situation. Each side maintains its own version of the map situation and records. Such games are used to test plans and for training. (There is a three-board form of chess on the same principle, called *Kriegspiel.)*

Which aspects of simulations discussed in the text are present in this game and which are missing? What are the advantages and disadvantages of this procedure? Compare this with a "computerized war game" in which the umpire group and the effective map are replaced by data processing equipment.

2.10. Describe an executive program suitable for (a) a critical event simulation, (b) a fixed time increment simulation, (c) a "track while scan" radar computer program, (d) a business record processor as in Exercise 1.8.

2.11. The executive program must choose the next module on the basis of information available in the computer and choice criteria. Analyze the forms this information can take and the possible executive programs.

2.12. A cathode-ray-tube display is either static or gives the impression of motion by a succession of frames, each of which is a complete static picture. But such a static picture is actually generated by the trace of a fast-moving spot whose x and y coordinates on the face of the tube and intensity are functions of the time. Describe the various ways in which a static picture can be analyzed into elements that have a precise numerical description and can be represented on a cathode-ray tube. A complex of such elements can then be represented in the computer as a "stored image" corresponding to a complete cathode-ray-tube display.

Many procedures have been developed to represent situations of various kinds by a two-dimensional display or by some combination of such displays. The corresponding stored images can be developed and manipulated in a consistent manner by the data processing system. Communication with people professionally involved has been well developed for such computer systems, and many operations that previously had required a great deal of human effort and skill are readily available as computer output.

Thus, one has a concept, analogous to language, with, however, a greater potential for simulation of both spatial relations and temporal developments. Discuss this concept relative to the design of machinery, architecture, and surgery. Include the possibility of automatic realization as represented, for example, by computer-controlled machine tools.

2.13. Discuss procedures for obtaining the characteristic roots of a symmetric matrix; also, methods for obtaining the largest root or the least in absolute value.

2.14. In view of roundoff, what does it mean to say that a set of variables satisfies an algebraic equation in the computer? Consider the three possibilities, single precision, double precision, and floating point. Given a nonsingular system of simultaneous linear equations and a set of values for an unknown that satisfy the equations in the computer, what can be said about the accuracy of the solution in each of the above possibilities? Discuss iterative procedures for solving simultaneous linear equations.

2.15. Three equations $F_i(x_1, x_2, x_3, t) = 0$, $i = 1, 2, 3$, determine x_1, x_2, x_3 as functions of t. Describe a process for obtaining x_1, x_2, x_3 corresponding to a stepped sequence of values of t. How can one establish the stability and accuracy of such a process?

2.16. Consider a system of n ordinary differential equations on n unknowns. How does a solution, i.e., a set of n functions, depend on an error in the values of the initial conditions, assuming the error is quite small? Show how the coefficients of the errors can be obtained as functions of the independent variable by solving systems of $2n$ ordinary differential equations on $2n$ unknowns.

2.17. In the step-by-step integration of a system of ordinary differential equations, one can regard each step as yielding the correct advance values from the previous values plus an error that can be regarded as an error in the initial conditions for the rest of the computation. This can be utilized to express the total error after a number of steps in terms of the error at each step. Various integration procedures yield step errors proportional to a power of the step size greater than one. Discuss the convergence of such approximations to the solution as the step size diminishes, ignoring the effect of roundoff. How can this be used to estimate the error in integration procedures as a function of step size? What is the effect of changing the scale of the dependent variable? What is the effect of roundoff?

2.18. A step-by-step integration process replaces a system of n differential equations by a system of difference equations. Discuss the stability of this process.

2.19. There are a number of procedures for obtaining solutions of linear partial differential equations with appropriate boundary conditions. Survey the applications of (a) separation of variables, (b) Green's function, (c) methods of characteristics, (d) relaxation methods, (e) finite difference procedures for parabolic and hyperbolic equations.

2.20. Discuss logically structured procedures for debugging a simulation program.

2.21. Training equipment based on simulation is used for many purposes, such as submarine crew training, carrier aircraft landing, astronaut docking maneuvers, and command control of supertankers. What would be appropriate equipment block diagrams, information flow charts, and integration plans in each case?

2.22. What professional training would be appropriate for simulation analysts?

2.23. The thermodynamic state of substances is important for the simulations of steam engines, internal combustion engines, internal ballistics of guns, explosions, and shock waves. An associated concept is that of ignition or the propagation of flames. What are the names and definitions of the thermodynamic and ignition variables that appear in these simulations?

2.24. Obtain the block diagram, math model, and flow chart for any system in the following exercises from Chapter 1 that interests you—Exercise 4, 10, 11, 16, 20, 21, 22, 23, 24, 26.

2.25. Investigate the procedures used to simulate optical devices, in particular, telescopes, microscopes, binoculars, range finders, cameras for photography, television cameras, sniperscopes, and the eye. How is ray tracing related to the lens formula?

2.26. A recent development has been the formation of three-dimensional images by holography. Investigate the mathematical basis of this concept.

2.27. Obtain block diagrams, math models, and flow charts for a pocket-size electronic calculator. (See McWhorter.[4])

References

1. Bush, V., Instrumental Analysis, *Amer. Math. Soc. Bull.* **42**, 649–669 (1936).
2. *Commemorative Technical Volume*, U.S. Naval Training Devices Center, Orlando, Florida (1971).
3. Connolly, Mark E., Simulations of Aircraft, Report 7591-R-1, February 15, 1958, Servomechanisms Laboratory, Massachusetts Institute of Technology, Department of Electrical Engineering, Cambridge, Massachusetts.
4. McWhorter, E. W., *Sci. Amer.*, **234**, 88–98 (March, 1976).
5. Murray, F. J., *Mathematical Machines*, Vol. 2, Columbia University Press, New York (1961).
6. W. Thomson, Mechanical Integration of the General Linear Differential Equation of Any Order with Variable Coefficients, *Proc. Royal Soc. (London)*, **24**, 271–275 (1875–76).
7. W. Thomson, Mechanical Integration of Linear Differential Equations of the Second Order, *Proc. Royal Soc. (London)*, **24**, 269–271 (1875–76).

3

Understanding and Mathematics

3.1. Experience and Understanding

We will now consider a general framework for "understanding." As individuals we deal with an environment that affects us either favorably or unfavorably, and our continued existence depends on our interaction with it. Experience is this process of dealing with the environment. Experience is certainly continuous, but we consider that we can resolve it into unit sub-procedures in which we deal with a "situation" and use our understanding to guide our actions to produce a favorable result. Clearly this understanding is based on past experience or on learning, which corresponds to the experience of others. Preceding experience must, therefore, be structured into patterns that we can recognize in the situation we are dealing with.

Dealing with such "units" of experience occurs on many levels of complexity, from our reactions to situations of immediate danger through many types of problems of everyday living, to the simulations we discussed in the previous chapter. These levels certainly differ in the extent to which we are aware of the structure of the process, and our interest is with those levels in which the relation of understanding and experience is explicit to the extent that communication between individuals and mutual action is possible. Communication and joint action also imply mutual understanding, which must include a common structuring of past experience. Part of this common structuring is educational in origin, part of it is the result of communicated individual experience, part of it is due to mutual experience, and there may be important elements that are innate or hereditary.

In order to associate patterns of past experience with the current situation, understanding must function in two complementary ways. As an integral part of the awareness process it produces in our consciousness a conceptual analog of the situation and it also produces experience patterns associated with the concepts involved, thus permitting us to explore, in our imagination, possible experiences corresponding to optional actions on our part.

The conceptual analog of the situation appears in terms of concepts that permit the formulation of experience patterns. For communication purposes, these concepts must be associated with symbols; for example, names and mutual understanding must permit us to associate the symbols and concepts. For each individual to have a correct version of these concepts, there must be a community sharing of experience and association of the symbols with the appropriate experience, and this is the function of education. The formulation of concepts may represent an innate tendency of our minds, but even if this is so, communication and mutual interaction of individuals certainly seem necessary in order to insure the similar nature of a concept in different individuals.

Examples of such concepts are represented by the forms of classical logic; for instance, the terms "man," "dog," and "triangle" represent a division of individual objects into classes we can readily recognize. Each of these classes is associated with a specific and complex combination of experience patterns. Reasoning associates this combination with the objects involved. We recognize that Socrates is a man and we know that men are mortal and we infer that Socrates will die. We know that dogs bite, meter maids give parking tickets, and 13 is an unlucky number.

This conceptual division into classes has certain general properties associated with the idea of class or set itself and with the identity of objects. The inclusion relation has certain properties, as does the notion of being a member of a class. These constitute an experience pattern called "logic," and this pattern is widely applicable. The terms "and," "or," "implies," "all," and "there is one such that" refer to logic. Language permits this conceptual experience to be shared and common agreement. It also permits one to check the conceptual development against the generally accepted characteristics of the experience patterns.

Thus, we have a formulation of experience, or at least part of it, in terms of objects and classes. Since this experience pattern is expressed in terms of symbols and language, it can be considered to be abstracted from the specific realization and becomes an intellectual experience itself. When the basis of such an intellectual experience is precisely described relative to the symbolic representation, the intellectual experience can be shared by many individuals and effectively utilized to structure experience.

"Logic" in the above sense is often considered part of an extended intellectual system that claims to represent all experience patterns. Such an extension is called "philosophy," but there is no universal agreement on what a correct philosophy is. "Logic" itself can be considered simply as a format for experience. This format is particularly valuable when applicable.

Mathematics consists of more general abstracted experience patterns, which include the logical pattern. The simplest of these and the one that has the widest pattern is that which includes counting, the natural numbers, and the notions of objects and sets and is usually referred to as "elementary arithmetic." As in the simpler case of logic, an area of mathematics is an intellectual experience associated with symbols and corresponding to an abstracted complex of experience patterns. The symbolic representation permits it to be shared and verified. To be useful as a experience format, it must be free from contradiction. Whether the pattern provides an effective conceptual structure for any specific situation must be verified by experience.

3.2. Unit Experience

Let us now describe by means of a flow diagram (Figure 3.1) a conceivable experience unit of a relatively simple nature. The experience units of interest to us are on a more sophisticated level, but we will show how this level can be obtained by a sequence of refinements. The nature of these refinements will be highly significant for us.

The flow diagram represents the sequence of actions of an individual. The individual directs his senses to observe a situation. This process yields information that is associated with available concepts to produce a conceptual analog of the situation. However, this procedure is not a unidirectional flow, but is interactive. The information suggests concepts that yield directions for additional observation. The diagram does not try to represent the various readjustments that are part of this process. Direction is indicated by an encircled D.

Available experience patterns are applied to the conceptual analog in order to consider possible actions. The logic patterns involving objects and classes are frequently useful to structure the intellectual development, but this is also true of the more general patterns of mathematics, especially arithmetic. In this imaginative development one also assumes various possible actions—one's options. It may also be true that the experience patterns available may not permit one to decide on a course of action, but the major objective is precisely such a choice.

If a decision is made and action is taken relative to the situation, this action results in certain effects on the individual. These effects are compared with those anticipated from the reasoning. If these are in agreement we consider the understanding involved in our conceptual analog that is the reason-

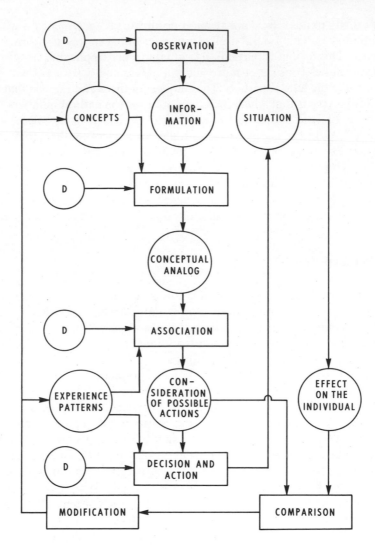

Figure 3.1. Interaction of understanding and experience.

ing, concepts, and experience patterns as satisfactory. Normally this is a very important element in our sense of security. If there is a disagreement we feel an urgent need to correct our conceptual analog of the system by further observation, and if our awareness seems satisfactory we must check our reasoning and if necessary our concepts and experience patterns. If it appears that these must be adjusted, then additional experience usually is required corresponding to the logical exploration needed to establish the new experience patterns.

Notice that there is the possibility that the concepts and experience patterns may be modified or adjusted. If we consider the combination of concepts and experience patterns as a body of knowledge, then there are various criteria for the validity of such knowledge that can be applied:

(a) Correct guidance is always obtained.

(b) Correct guidance is obtained when explicit tests are made.

(c) There is a sequence of modifications each of which yields correct guidance for a more inclusive set of circumstances.

(d) Satisfactory guidance is obtained in the sense that the cases of incorrect guidance are considered unimportant.

(e) One can introduce modifications on a given set so that correct guidance can be obtained in any case of interest.

Clearly this list can be extended, but it does illustrate how much latitude exists in the concept of a satisfactory body of knowledge.

There are various philosophical points of view that assume that certain aspects or blocks in this diagram generate all the others. For example, there is the point of view that the line joining "situation" and "observation" is the critical origin of the diagram that starts all else. Other points of view will begin with either the "concepts" box or the "experience pattern" box and assume that these are generative. We will simply assume that at any time, experience will at least contain the indicated complications, that all the elements will evolve in time, and if some one wishes to understand the overall development or even its present nature he must look for as much of the past history as is available.

For an introduction to various philosophical approaches, one can consult Joad.[4] The Aristotelian "universe of individuals" is described by de Wulf.[2]

3.3. The Exact Sciences

We now consider various refinements in the experience scenarios given in the preceding section that correspond to the cases of primary interest to us. It is convenient to discuss these refinements in terms of a block diagram in which certain procedures are lumped together under rather obvious headings (Figure 3.2). Awareness yields a conceptual analog in our mind that must be matched with structured experience to permit reasoning concerning outcomes. Understanding corresponds to the combination of concepts and experience patterns that is the basis of reasoning. We are particularly inter-

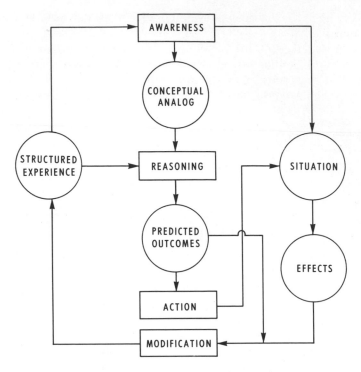

Figure 3.2. Simplified version of interaction.

ested in the case where the conceptual experience patterns include those that are "mathematical."

A mathematical intellectual experience can be considered constructive inasmuch as the various patterns of experience can be combined into larger complexes, and such a construction has a symbolic representation. The usual mathematical operations are steps or units elements in these constructions. Thus, the imaginative development of one person can be communicated and also symbolically checked by himself and others. Hence, mathematics is particularly suitable for planning and cooperative activity.

The constructive nature of the operational procedures permits considerable imaginative exploration in any situation that can be linked to mathematical concepts. The extensive development of the mathematical operational procedures themselves is available and is highly useful. We have seen in simulations the production of mathematically based time histories to establish possibilities. Thus, when mathematics can be used it greatly extends the experience patterns that are available from logic.

Many sciences have such a mathematical aspect. A science is an overall complex of methods for dealing with a milieu of experience. In the "exact" sciences, there is a central mathematical formulation that permits a purely mathematical investigation of the situations of the experience milieus. Thus, a given situation of this type can be considered to have a math model, which is obtained by an analysis of the situation in terms of the conceptual experience patterns of the science.

The analysis of the situation can be considered to be the equivalent of setting up an influence block diagram in which the blocks correspond to the concepts of the science. In the exact sciences the concepts are associated with precisely disciplined procedures, for example, measurements, whose outcome is mathematical. Typical examples are lengths, volumes, weights, angles, temperatures, and electric·currents. Mathematical relations between these outcomes that yield satisfactory predictions for· dealing with the milieus constitute the math model. One can have functional relationships or more subtle ones involving derivatives or probabilities. Mathematical procedures are available for reasoning based on the known mathematical relations.

Sciences are usually described as "bodies of knowledge," and there is an implicit separation of the "theory" or mathematical formulation of the structured experience from the rest of the understanding. Furthermore, this theory or mathematical formulation is considered to contain all the "knowledge" of the science. But such a theory must be supplemented by a procedural interpretation of the concepts. The applied mathematician must appreciate that "scientific understanding" refers to the complete procedure for dealing with a milieu of experience including the actual operational procedures. Changes and improvements in the latter are often the essential aspect of a scientific advance.

This is important also for the following. By definition, the reasoning associated with an exact science can be expressed in symbolic mathematical form. But normally, the reasoning is not confined conceptually to purely mathematical procedures. A wider range of conceptual reasoning is used based on the additional concepts and experience relations of the science itself. The mathematician must learn that this is quite appropriate. The concept of a "particle" in physics has a set of mathematical variables that specify it precisely, and yet conceptually one prefers to reason with it as an idea in its own right.

Our distinction between the mathematical model and the rest of the understanding is relatively modern and corresponds to a belief in the conventional and inventive characteristics of mathematics. The natural philosophers of the eighteenth century believed that both mathematics and the various scientific logical developments were part of an axiomatic structure by which we comprehended nature.

3.4. Scientific Understanding

As "understanding" the conceptual aspects and mathematical theory of a science is subject to the requirement of being a continuing satisfactory match to a milieu of experience. This match is usually thought of as involving the maximum available exploratory testing, i.e., "experimenting," but it must also include the procedures of applied technology. The intellectual development of a science is part of the broad experience by which a culture advances and yields advanced technologies. On the other hand, technical improvements expand the available areas to which the scientific understanding is applicable and are incorporated into the operational aspects of scientific concepts. The expansion of the milieu and the improvement in conceptual procedures frequently require essential changes in the intellectual aspects of science.

The mathematical formulations of technical procedures have contributed very significantly to the spread of applications, as for example, in circuit theory, fluid flow, and statistics. Usually the basis of a technological advance is some phenomenon for which a scientific basis has been developed in the form of quantititive mathematical relations. The latter is then adapted to the needs of the technical situation and becomes a part of the engineering repertoire. This advance of technology due to scientific developments is probably very familiar to the reader.

On the other hand, many fundamental improvements in scientific understanding have resulted from advances in instrumentation, and these advances were due to technological advances. Such developments are of course well known. The use of optical instruments was a critical advance in many sciences—the microscope in biology, the telescope in astronomy, the spectroscope in physics and chemistry. Before the telescope, ancient instruments permitted the original Ptolemaic description of the solar system, but improvements in these instruments were adequate to support Kepler's laws and their Newtonian consequences.

An educated person should certainly be aware of this complex interaction of science and culture. One aspect is rather important for the applied mathematician. He must appreciate that scientific concepts are based on disciplined technical procedures and that the consequences of a mathematical theory must be interpreted in terms of these concepts. This applied in particular to the "data" that must be used in technical applications. The experimental context of such data must be carefully understood and may correspond to essential limitations. The precise procedural equivalent of various notions is usually very interesting, especially from the point of view of the manner in which the basic procedure has to be supplemented. Useful information about the length of a bar of metal may have to include the temperature and tension. The measurement process becomes increasingly complex as one deals with the distance concept in machine parts or variations on the surface of a telescopic mirror. On the other hand, the distance to heavenly bodies requires a pyramid of procedures. In this case it is clear that the simplicity and computational aspects of the mathematical concept have been retained at a considerable expense of complexity for the experience patterns.

This discussion indicates the problems associated with the awareness process when scientific analysis is required. The mathematics to be used may be available in standard form, but not the data that specify the situation. Experimental information must always be considered in the context of the actual procedures, and these procedures may be quite disparate, even when the same expression is used and has the same mathematical significance. For example, the expression 'speed of sound in sea water" has to be interpreted very carefully to insure that one knows what parameters (such as temperature, density, depth, and salinity) have been taken into account and how they were considered. "Scientific facts" or "scientific data" are significant only in a framework of scientific understanding. Experience has shown that theorctical unifications based on purely conceptual or mathematical identity can lead to false results.

3.5. Logic and Arithmetic

We frequently use the experience pattern of objects, classes, and natural numbers to structure situations. We make up shopping lists, laundry lists, lists of our possessions, and lists of our investments, and we make numerical comparisons and evaluations. Thus, arithmetic is an extension of elementary logic. One uses set notions to divide objects into various categories and express the relation between categories. Many aspects of the situation are expressed in terms of counting and the arithmetical operations. For example, the total cost of a shopping list is evaluated by these means. Thus, elementary arithmetic partakes of the nature of logic and deals with the same general experience patterns.

Agreement on these experience patterns relative to the counting process and its outcome, numbers, is of great practical importance. Thus, one must have agreement on the symbols for numbers and the effect of the arithmetic operations. The abstract conceptual situation is such that specific experience can be used to establish general relations, i.e., since two apples and two apples make four apples, we know that $2 + 2 = 4$. But this also means that one can use *ad hoc* simple examples such as set of marks to establish arithmetic relations by applying the appropriate set of operations; for example, "six times seven is forty-two" can be shown by taking six sets of seven marks and counting them.

But in all cases of practical interest, this unsophisticated approach must be supplemented by using the notion of aggregation, or forming sets, for counting purposes. This is extremely familiar and results in our decimal notation and the abacus. Sets are represented on the abacus by a uniform system of aggregation, and the use of the abacus corresponds to the effect of the basic set operations on counting. The decimal notation corresponds to a somewhat greater use of symbolism but has essentially the same character.

It is interesting to notice the analog character of the abacus. Thus, if a situation is reflected into prices, the value relations can be represented on the abacus.

3.6. Algebra

In planning, problems occur that are solved by a sequence of arithmetical operations called an algorithm. Our earliest known historical examples of arithmetic are in this form. A specific problem is stated and an arithmetic procedure is given to solve it in terms of the numbers in the problem. The reader is supposed to recognize that this pattern of operations can be applied to a class of problems.

Abstracting this experience into effective symbolic form occurred in historical times and led to what is now termed "algebra." The symbolism includes representations of numbers either unknown or unspecified and operations. Relations that are consequences of the arithmetic operations are expressed by equations that can be manipulated to yield other relations. Such manipulations reflect basic set relations and cannot be established by observing their effect on specific numerical relations. They must be referred to the set relations on which arithmetic relations are based. They constitute experience patterns abstracted from the basic arithmetic ones.

In modern mathematics, there are notations for sets and sets operations in which characteristic properties can be expressed. The arithmetic operations are defined in terms of set operations. For example, the sum corresponds to the union of disjoint sets and the product of numbers correspond to the cross product of sets or the set of pairs. Thus if n_i is the cardinal number of the set, A_i, $i = 1, 2$, the sum corresponds to $A_1 \cup A_2$ and $n_1 n_2$ corresponds to $A_1 \times A_2$. The properties of the arithmetic operations, for example, commutativity, associativity, and distributivity, follow from those of the set operations.

This notation corresponds to a further level of abstraction. A number like 3 or 7 corresponds to an abstraction from the counting process, which itself must be considered as a relatively general experience pattern applicable to sets. Algebra deals with an abstraction process for experience patterns involving numbers in general and their properties. These properties are also associated with general relations between sets. These successive layers of abstraction from experience patterns also illustrate the role of symbolism in such abstractions.

3.7. Axiomatic Developments

Numbers represent abstractions from experience patterns, but we now regard them as objects subject to the set operations that we have applied previously to the objects of direct experience. In particular we have a set of all natural numbers and we can consider subsets and set operations such as union, intersection, formation of a set of pairs, and consequence (the notion of one-to-one correspondence). This is a considerable expansion of mental experience possibilities. We do not deal with infinite sets of objects, for example, in normal experience.

Since the set of numbers constitutes an independent set of objects, we can structure the associated experience axiomatically. This means that certain characteristic properties and relations are postulated and all other properties and relations are deduced from these characteristics by logical operations, that is, by set properties and constructions.

The standard way in which this is done for the natural number is with the Peano axioms (see E. Landau, *Foundations of Analysis*[5]). The set, N, of natural numbers is characterized by two set-theoretic properties and a specific correspondence, $n \sim n'$. We have

(1) $1 \in N$
(2) $n \in N \rightarrow (\exists n')(n' \in N)$
(3) $(n)(n' \neq 1)$
(4) $n' = m' \rightarrow n = m$
(5) $(M \subset N) \cap (1 \in M) \cap [(n \in M) \rightarrow (n' \in M)] \rightarrow M = N.$

We have used the notation of graduate mathematics for set relations. While Axioms 1–4 deal with N and its elements, Axiom 5 is a statement about any subset M of N and is a conceptual expansion.

On the basis of these five axioms one can construct functions of two variables, "plus" and "times," which have the appropriate properties, and also a notion of "less than." We have a one-to-one mapping of the natural numbers abstracted from immediate experience into this axiomatically defined set. The logical development based on the postulates is essentially different from the arguments based on abstraction.

Having defined operations such as addition and multiplication for the natural numbers and having established their properties one can consider a set of abstract objects subject to these operations. This set of objects is then "defined axiomatically" by the existence and properties of the operations. There is considerable leeway in choosing the properties associated with the operations, and this choice leads to various "abstract algebras."

An axiomatic discussion presents a logical structure that is quite valuable in obtaining mutual agreement. The alternative is multiple *ad hoc* arguments. The axiomatic logical structure is independent of any preceding abstraction process and presents its own criteria of rigor.

Axiomatic discussions are also applicable to areas of experience not associated with discrete objects. Measurements deal with procedures for comparing magnitudes such as line segments or surface areas. The naive approach to comparing two magnitudes such as a pair of line segments would be to express both as multiples of a common unit and refer the situation to fractions or pairs of integers. But even in ancient times this was known to be inadequate, and Euclid's *Elements* (see Heath[3] and Morrow[6]) contains a more sophisticated approach.

Geometry deals with a system of interrelated ideal objects—points, lines, planes—that clearly corresponds to an abstraction from spatial experience. This system is described axiomatically with the corresponding logical structure. In Euclid, arithmetic and algebra were incorporated into geometry by axioms.

Elementary analysis (see Landau[5]) can also be given in terms of an axiomatic definition of real numbers and certain set ideas. In general a mathematical axiomatic discussion deals with a set of ideal objects subject to certain experience patterns abstracted from some area of previous experience. The ideal objects and the relations between them have a symbolic repre entation that can be used to insure that the imaginative development is consistent with the specified experience patterns and can also be used for communication. Clearly these imaginative developments can be piled one upon the other, i.e., the experience patterns of one can be the basic axioms of a higher development. From our present point of view, the associated symbolic construction is a representation of a purely intellectual development not the development itself.

3.8. Analysis

The natural philosophers of the eighteenth century expanded the experience areas involved in the axiomatic development to include motion, gravity, electricity, magnetism, temperature, fluid flow, and the elastic properties of matter. But this was done at a price. The contact of the intellectual formulation with experience in these new areas was by experiment. However, it was clear that one could separate out a central core of mathematical procedure, not dependent on experiment, that was used in reasoning and prediction by computation. This central core was called analysis and it did represent conceptual experience adequate to provide math models for natural philosophy.

In modern terms analysis corresponds to the theory of functions of a number of real or complex variables. In its original form it was not dependent on experiment, but it did involve computational procedures using infinite sets of quantities that had meaning only because of *ad hoc* intuitive appeals. Furthermore, analysis did not stand by itself as an axiomatic development but was supported by intuitive appeals to the original unabstracted situation. The notion of function and limit, for example, were based on the idea of motion.

These logical difficulties were understood. In addition the discovery of non-Euclidean geometry showed the imaginative and inventive character of mathematics. Thus, it seemed desirable to set up analysis as an axiomatic structure in which the constructive procedures utilized only set-theoretic methods.

The resulting axiomatic development begins with the Peano structure for the natural numbers and establishes the arithmetic properties and the notion of "less than." The positive rational numbers are defined as sets of equivalent pairs of natural numbers. Two pairs (m_1, n_1), (m_2, n_2) are equivalent if $m_1 n_2 = m_2 n_1$. The algebraic properties of positive rational numbers include the group property for multiplication. The extension of these obtained by adjoining 0 and -1 algebraically is a field with an ordering relation less than. The real numbers are defined in terms of set constructions by the Dedekind cut process. The reader is undoubtedly aware of the set-theoretic constructions that correspond to the notions of complex variable, functions, limits, and analytic geometry.

Thus, analysis is established on a set-theoretic basis starting with the natural numbers. Previous mathematical developments are subsumed by mappings onto parts of analysis in such a way that "intuitive elements" are replaced by set-theoretic arguments. The exact sciences have math models that represent in this sense both geometrical and physical concepts. Quantitative aspects of reasoning are referred back ultimately to integers.

Even the Peano axioms can be replaced by set theoretic constructions. Suppose that there is an object, a, that is not a set and we consider a set $\mathscr{A}$ of sets A that has the following properties:

(i) $A_1 = \{a\} \in \mathscr{A}$;
(ii) $A \in \mathscr{A} \rightarrow \{A\} \in \mathscr{A}$;

(iii) $(B \in \mathscr{A}) \cap (A \in B) \to A \neq B$;

(iv) $(\mathscr{B} \subset \mathscr{A}) \cap (A_1 \in \mathscr{B}) \cap [(A \in \mathscr{B}) \to (\{A\} \in \mathscr{B})] \to \mathscr{A} = \mathscr{B}$.

Intuitively, the set $\mathscr{A}$ consists of the sets

$$\{a\}, \quad \{\{a\}\}, \quad \{\{\{a\}\}\}, \ldots,$$

and there is a one-to-one correspondence with the natural numbers 1, 2, 3, ... However, we can readily show formally that $\mathscr{A}$ satisfies the Peano axioms. We sketch the discussion.

Clearly $1 \sim \{a\}$, and if $A \sim n$, $\{A\} \sim n'$ and we have Axioms 1, 2, and 5. To show Axiom 3 (i.e., $n' \neq 1$) and Axiom 4 (i.e., $n' = m' \to n = m$), we show the following theorem:

$$(A \in \mathscr{A}) \cap (b \in A) \cap (c \in A) \to b = c.$$

Let

$$\mathscr{B} = \{A : (A \in \mathscr{A}) \cap [(b \in A) \cap (c \in A) \to b = c]\}.$$

But the definitions immediately yield $A_1 \in \mathscr{B}$ and $A \in \mathscr{B}$ implies $\{A\} \in \mathscr{B}$. Thus, (iv) above yields $\mathscr{A} = \mathscr{B}$, and hence the theorem. The Peano Axioms 3 and 4 follow readily from the theorem.

Thus, all analysis can be expressed in set-theoretic terms, assuming the existence of a single object. It is "logical" in the sense that the conceptual experience consists of "constucting sets," a mental process in which objects that themselves may be mental are associated into a set.

This set-theoretic analysis also leads to axiomatic developments. Axiom patterns can be abstracted from various parts of set theoretic analysis, with, of course, the set theoretic concepts themselves. The logical structure consists of arguments, symbolically described, corresponding to the abstracted patterns and set theory. This yields a very extensive mathematics, including abstract algebras, topologies, and linear spaces, i.e., graduate mathematics. These, of course, frequently parallel previous axiomatic developments.

The basic drive behind this development was the demand for rigor. It is now widely accepted that set theoretic arguments are logically satisfactory. For axiomatic structures, in general, which are considered formats for experience rather than experience itself, freedom from contradiction is certainly satisfactory (see Landau[5]).

3.9. Modern Formal Logic

There are those who object to the characterization of mathematics as an "intellectual development" and who wish to assign all essential significance to the symbolic procedure. Modern "formal logic" is based on the principle that there are certain formats of words and statements that are true. For example, "*A* included in *B* and *C* included in *A* implies *C* is included in *B*." The concern of "formal logic" as represented, say, in the *Principia Mathematica* of Whitehead and Russell,[10] is to set up an axiomatic development of these formats. The ingenious system of symbols of formal logic is widely used.

This principle of formal logic involves the idea that the "truth" is associated with symbolic forms and procedures. Of course, the reason one is willing to accept such syllogisms as the above is that one can imagine classes of objects corresponding to certain denoted properties. Mathematics is usually presented in set terms with imaginative set constructions. But this principle requires that all such intuitive or imaginative elements be replaced by explicit symbolic formulations. The corresponding notation is familiar to the reader, for example, $(x)(P(x))$, $(\exists x)(P(x))$, and $\{x: P(x)\}$.

Thus, the usual development of analysis can be represented in this notation, i.e., in appropriate symbolic form. But this requires a postulational justification for the various set constructions, and the obvious procedure is to set up a set of rules governing the construction of the symbols for sets. For example, if one has a property for which one can express the fact that this property holds for x by the symbol $P(x)$, then one has the set $\{x: P(x)\}$.

In the present context, this is equivalent to an axiomatic development of set theory. The postulates are the construction rules for symbols for sets. In the last century, Frege proposed such an axiomatic formulation as a fitting climax to the effort, sustained over two hundred years, to obtain a precise rigorous mathematics.

The result was a catastrophe. Bertrand Russell showed that the axioms led almost immediately to a contradiction. Let $\sim$ stand for the denial of a statement. Consider the set $\sigma = \{x: \sim(x \in x)\}$, i.e., the set of those constructs that are not members of themselves. Then $\sigma \in \sigma$ implies $\sim(\sigma \in \sigma)$ and $\sim(\sigma \in \sigma)$ implies $\sigma \in \sigma$.

In general, the idea of applying set concepts to the set of sets leads to difficulty. The construction procedure must be much more sophisticated. Russell proposed a hierarchal structure that was not subject to the given difficulty and other construction procedures have been proposed. But this also developed problems of undecidability.

These difficulties led to an analysis of the symbolic procedures of mathematics. In this analysis mathematical concepts are used on a frankly conceptual basis with abstractions from the symbolic procedures. As a consequence one has a mathematical treatment of general questions concerning the structure of mathematics such as consistency, categoricalness, and decidability. This procedure is called "metalogic."

It is not clear that one can obtain every aspect of mathematics by symbolic procedures alone without sharing conceptual experience. The symbolism of mathematics is designed for communication and is considered successful if the designated intellectual experience is shared. The requirements of formal logic constitute an additional demand on the symbolism that it be associated with "truth" in the Platonic sense.

Thus, the symbolism has the character of the higher unchanging logic that Proclus sought. Proclus objected to mathematics because its truth is conditional, i.e., the theorems require hypotheses and the conclusion is valid only when the hypothesis is satisfied. But the "true" statements in a symbolic formal logic are valid unconditionally. Thus, $(p{\rightarrow}q)\cap(q{\rightarrow}r){\rightarrow}(p{\rightarrow}r)$ is always valid. It is not evident that this extra requirement will be universally accepted or how it can be satisfied. On the other hand this requirement does not appear necessary for applied mathematics.

A formal logic development is axiomatic and can be considered to be the construction of a sequence of symbolic expressions. Constructive procedures of this type can be subject to mathematical analysis in order to deal with questions of consistency, independence, and decidability. The experience patterns of formal logic can be abstracted and supplemented by set-theoretic analysis. This yields a sophisticated mathematics of great interest in itself that can be regarded as a conceptual format for experience. Thus, one can consider the modern theory of formal logic as mathematics without making the assumption that all logical experience must conform to it.

Mathematics is frequently described as a language, and it is worthwhile to make this comparison. In both cases, one has symbolic communication between people relative to a common background of experience and the possibility of an imaginative projection of experience. Of course, in mathematics the symbolism is precise enough so that the usual ambiguities of language are eliminated. But communication is only one aspect of mathematics. Mathematics also involves an intellectual development consisting of abstracted patterns of experience. This development is the subject for the communication in mathematics and is more central than the communication aspect. Mathematics is equivalent to a language plus a literature.

For further discussion of the material in this section, see Church,[1] Morrow,[6] Styazhkin,[8] Takeuti and Zaring,[9] and Whitehead and Russell.[10]

3.10. Pure and Applied Mathematics

It is true that there are certain difficulties in applying set concepts to the collection of all sets. However, in regard to analysis there is considerable confidence that a satisfactory symbolic representation of the set-theoretic structure exists. Thus, starting with the existence of a single object, the mathematician can construct the whole of mathematics, using no imagined procedure except the forming of sets.

The mathematician exclaims "*C'est moi*," and it is clear that mathematics is an exclusively intellectual development. This is indeed the way mathematics is presented in our graduate courses. The mathematician needs only mathematics in his research and lectures. Previous forms of mathematics are incorporated into the present one in a new rigorous form, and this independence permits a wealth of development, which is of great interest and offers the greatest possible freedom for scientific exploration. The applied mathematician is welcome to use this logical development, but this use as such is not considered to contribute to the fundamental mathematical structure.

But this independence definitely involves a certain isolation. Most areas of applications involve, conceptually, earlier forms of mathematics and are still taught in this way. The user develops a familiarity with some such development and may be unaware of the corresponding "pure" or modern mathematics. His mathematics may correspond to the early nineteenth century or to one of the two forms of the eighteenth century. These, of course, have been subsumed into modern mathematics, but the reward to the user for mastering a new and demanding intellectual discipline is frequently just the assurance of rigor and not necessarily any increase in capability.

Indeed, the increase in capability in the sense of obtaining new computational procedures may occur outside the framework of accepted mathematics, as for instance, the Heaviside operational calculus or the Dirac delta function. Mathematics has been expanded to include a justification for these. This is an intellectual triumph and represents some increase in applicability, but it also shows that intellectual restrictions are basically unjustified.

One must then anticipate the possibility that the user of applied mathematics will function intellectually in a way that is independent of "pure" mathematics, while the mathematician will learn his mathematics in the latter framework. Technical developments frequently make it desirable to tap the rich vein of available mathematics. In most applied areas there is a relatively standard mathematics, which is normally adequate, but the explosive development of technology may require an expansion in the mathematics used. The use of differential equations was tremendously expanded by the availability of automatic computation, but this expansion required a new theoretical base in existence, and uniqueness results from pure mathematics.

The applied mathematician must make such contributions, and he must also involve his mathematics in the general understanding of the situation. But this requires a somewhat different intellectual background than that for pure mathematics, and indeed the development of such an intellectual background can be a very satisfying professional activity. One must appreciate that the present conceptual structure of mathematics is the result of a historic process in the last hundred years, motivated by an overriding concern with logical precision. Previous phases of mathematics have been an integral part of what is termed the "scientific revolution" and were concerned with developing the power of mathematics. There are clear indications that initially the fascination of the conceptual development itself was the major element.

Our text contains an introduction to these matters but by no means a complete account. Our next objective will be to provide a historical framework in which the present development of the exact sciences and mathematics will appear in perspective. The pure mathematician who does not want mathematics to be considered to be in an isolated subculture may also find this of interest.

3.11. Vocational Aspects

It is desirable to summarize the significance of this chapter from the vocational point of view of the applied mathematician. The understanding of the situation that is the target of the effort is not a "body of knowledge" but a procedure for coping with the situation. Part of this procedure is the effective use of the math model and computation. The appropriate role for the applied mathematician is the responsibility for incorporating this effective use and computation into the overall procedure.

This requires an "understanding" of the basis for the math model in the sense that one must appreciate the experience patterns that correspond to the concepts that are used in handling the situation. Information on relevant past experience and desirable and undesirable results are expressed in these concepts. "Facts" do not exist independently of a conceptual framework of past experience.

The type of understanding described in the previous paragraph generally involves a considerable effort to comprehend technical reports, especially in regard to their quantitative significance. This may seem overemphasized, but there have been simulations in which the computed results were entirely irrelevant to the situation. In the early days of digital computation, an effort was made to function on the principle that the user group would prepare a mathematical formulation of the problem and a group associated with the computer would solve it. Presumably, this would lead to a most efficient use of the computer, but the total result was disastrous on occasion. The difficulties were compounded by a number of factors, but the major weakness was the limitation on the conceptual communication.

Communication must be in terms of the mathematical and technical procedures of the group, and the mathematician normally must exert considerable initiative to establish this. Certainly he cannot wait until someone else "expresses the problem in mathematical terms." The asset the mathematician has is the tremendous amount of mathematics available. But he will have to make the conceptual connection. This means he must develop a scientific understanding associated with his mathematics that is not readily available academically. He must develop an appreciation of the historical dimension in general, as well as in the specific situation he is dealing with.

Exercises

Term Project: (see p. 8): In your term project, describe the situations with which one deals, the concepts used in describing the situation, the experience patterns used for prediction and the mathematical formulation used for them, the conceptual development used for reasoning, the action options available, and the comparisons used for decisions.

3.1. For each of the experiences listed below, describe (i) the associated areas of experience, (ii) the possible favorable and unfavorable aspects, (iii) the experience patterns of actions and consequences that we take into account, (iv) the extent and general character of our awareness of these experience patterns, (v) the communication processes involved, and (vi) whether the situations involved are present, future, anticipated, obtainable, or avoidable.

(a) Taking a course at a university
(b) Employment
(c) Shopping in a supermarket
(d) Learning to drive a car with a manual shift
(e) Seeing a stop sign at the next intersection
(f) Seeing a state police car with a radar device on the highway ahead
(g) Paying one's annual federal income tax
(h) Watching a play or motion picture
(i) Playing chess or bridge
(j) Gambling with dice or betting on horse racing
(k) Participating in sports
(l) Reading
(m) Having thirteen guests for dinner

3.2. In discussing each area of Exercise 3.1, various nouns are used. In which cases do these nouns correspond to the names of concepts that are classical logical forms? What attributes associated with these forms are relevant to the discussion? In which cases is there a definition associated with these nouns that indicates a pattern of experience or an attribute? When does the definition permit a range of conceptual experience? What is your basis for understanding the form or definition?

3.3. Describe the different ways in which we become aware of situations of interest to us. What information is available to us without any action on our part, what information is due to involuntary actions on our part and what information results from deliberately directed actions on our part? Are these distinctions always clear cut?

3.4. In dealing with a situation do you list the possible actions you could take, i.e., your "options"? Do you think you should?

3.5. Have you ever had occasion to change your ideas about some aspect of experience? Can you describe the sequence of events that led to this change? Can you diagram the corresponding experience?

3.6. At the end of Section 3.1, there is a list of possible criteria for a body of knowledge. Expand and discuss this list and Pilate's famous question, "What is truth?"

3.7. List the comprehension concepts of geometry, physics, astronomy, chemistry, and biology. How does one learn the meaning of these concepts?

3.8. Describe the procedural definition of length in physics using the standard definition. Investigate the technological applications of this definition. What is the procedural definition of distance in astronomy? Is there a procedural equivalence between the concepts of length and distance?

3.9. If the result of a measurement is a real number, the choice of scale for the measurement is usually arbitrary. Describe dimension theory and its significant applications in aerodynamics, thermodynamics, and electromagnetism.

3.10. Various modern engineering procedures have been based on an analogy between electrical circuits and mechanical systems. What is the mathematical formulation corresponding to the concepts involved? What concepts are more readily available in one system than the other?

3.11. Descriptive biology is closely associated with certain concepts derivable as formal concepts of classical logic. Describe the development of the associated "classification" systems and the dependence of the recognition procedure on the point of view of the originators and the availability of techniques. Describe the associated diagrammatic representation. How was the conceptual structure affected by developments in physics, chemistry, and by the theory of evolution?

3.12. Describe the various milieus of experience associated with classical mechanics. How did this experience develop historically? How did the quantitative concepts develop and what are the now accepted mathematical relations involved? What are the present situations of technical interests? What are the mathematical procedures used for analysis and prediction? Answer the same questions for classical electrodynamics and thermodynamics.

3.13. The following mathematical topics are associated with various areas of applications.

(a)	Ordinary differential equations	(f)	Linear operators in Hilbert
(b)	Partial differential equations		space
(c)	Fourier series	(g)	Linear functionals
(d)	Fourier transforms	(h)	Integral equations
(e)	Probability distributions		

What are the comprehension concepts associated with the various applications? Name and describe the mathematical procedures used for analysis and prediction. Describe the formulation of the output information required and its relation to comprehension concepts.

3.14. Classify the concepts that appear in our dealing with matter in its solid form. In which cases are quantitative measures associated with these concepts? Notice that in certain areas of experience weight is considered a direct measure of quantity, but in elementary physics weight is simply a force, and quantity is associated with inertial mass.

3.15. Define the concepts "liquid" and "gas" in terms of experience. What is the relation to the notion of "solid"? How would one describe the general notion of matter? The sciences of physics and chemistry "explain the nature of matter." What does this mean in terms of the quantitative concepts discussed in this and the preceding exercise?

3.16. In modern physics an elementary particle "has both particle properties and wave properties." What is the experimental and mathematical meaning of this statement? What are the associated concepts?

3.17. Describe the experimental verification of Newton's law of gravity. It is usual to consider that Newton's law of gravity and the associated mechanics applies to a limited mileu within the area of general relativity. Describe this idea mathematically and describe the associated experimental verification. Do this for the similar relation between "geometric" and "physical" optics. Also for classical and quantum electrodynamics. Also for classical and quantum thermodynamics.

3.18. Diagram the relation between the various topics that are "exact sciences" in our sense in physics. Indicate how comprehension concepts structure the relations between them. Indicate how measurement procedures vary within these concepts and how geometric and space–time concepts are used to unify.

3.19. What is the relation between counting, keeping accounts, and arithmetic? What is the fundamental concern? What are the comprehension concepts and experience patterns used to translate various business activities into a form to which arithmetic is applicable? What are the action decisions resulting from this process and what are the desirable results? What is the meaning of the terms double entry bookkeeping, assets, debits, capital, purchase, sales, inventory, production, cost, cost accounting, gross and net return, and growth? What is the relation with arithmetic? What arithmetic checks are used?

3.20. Describe verbally the patterns of arithmetic experience that are expressed symbolically in algebra.

3.21. Geometry has been described as the idealization of spatial experience. Its name and certain historical evidence indicates that this spatial experience was surveying. What are the concepts used in surveying? In what sense are they idealized? The development of mathematics has added many more geometrical concepts. Describe them and their corresponding spatial experience.

3.22. Bertrand Russell in his essay,[7] "Mathematics and the metaphysicians," states, "Pure mathematics was discovered by Boole, in a work he called the 'Laws of Thought' (1854).... His book was in fact concerned with formal logic, and this is the same thing as mathematics." Outline the basis for Russell's statement by references to the literature. Is there a conceptual framework inherent in the establishment of formal logic? Russell also states that, "The propositions [of logic] can be put in a form to apply to anything." How does Russell prove that $2 + 2 = 4$ is a theorem of formal logic? How does this apply to four apples? Two apples and two pears? Milk? Bubbles? What concepts did you add to make the theorem applicable? Is the result always significant? What is the relation with the question: is the milk sour?

3.23. Under the influence of Newtonian particle physics, certain early theories of elasticity were based on the assumption that matter consists of "atoms" in certain arrangements with forces acting between them. In "mathematical elasticity," matter is supposed to be homogeneous but not isotropic, and concepts of stress and strain are defined. What common concepts permit a theoretical comparison and an experimental comparison of the two theories? The texts on elasticity insist that the first theory is "essentially different from modern atomic theories." Explain the difference in mathematical theory. What are the ranges of experience associated with this difference?

3.24. Describe the aspects of the following, which involve the construction of sets: (a) a topology on a set, (b) closure, (c) complementarity, (d) union, (e) intersection, (f) Baire sets, (g) filters, (h) maximal filters, (i) all subsets, (j) completion, (k) measurable sets, (l) sets of measure zero, (m) nonmeasurable sets, (n) metric spaces, (o) continuity, (p) Zermelo's axiom. When is the constructive aspect described axiomatically? When do you think it should be?

3.25. In the usual development of the theory of finite groups, what theorems from the theory of finite sets are used? Where are these proven? Give a complete axiomatic basis for the theory of finite groups. In the theory of a finite dimensional algebra over a field K, what structures from set theory are added to the properties of K? Give a complete axiomatic basis. Banach's original treatment of linear spaces was based on a sequence of "definitions" in each of which additional mathematical apparatus was introduced on what was initially a "linear vector space." Describe these spaces axiomatically. Banach also uses the concept of "transfinite induction." What are a complete set of axioms for this notion? How are these related to Zermelo's axiom and Zorn's lemma? What is the relation of Hilbert space to the spaces described by Banach?

3.26. An important initial part of the development of analysis is concerned with the properties of continuous functions of one real variable on a closed finite interval; these properties permit the definition of the Riemann integral. What property of a closed finite interval that is not valid for the set of all real numbers is crucial in this discussion? What set constructions and axioms are used in establishing it? What logical assumptions are involved in its use?

3.27. In linear space theory one is usually concerned with linear sets or closed linear sets, and the "element of" relation is replaced by an inclusive relation of a one-dimensional subset. The concepts of union, intersection, and element of for ordinary sets in general satisfy certain algebraic relations and constitute, therefore, a "Boolean algebra." The union of two linear sets is in general not a linear set. One can, however, obtain an algebraic structure associated with linear sets called "lattice theory." Describe the axioms of lattice theory. How are these applied to linear or closed linear sets? Contrast these with Boolean algebra. Obtain Boolean relations that fail in the lattice case. Boolean algebra is associated with elementary formal logic. What would be the result if one used lattice theory instead? What difference would it make if one used translates of linear sets instead of just the linear sets?

3.28. Give the formal symbolic arguments for the proofs in Section 3.8 of statements 1 and 2 concerning the set theoretic formulation of the natural numbers. Underline purely set-theoretic assumptions.

3.29. Formalize elementary number theory, starting with Peano's axioms and obtaining the theorem on unique factorization.

3.30. One important meta-type argument is Gödel's proof of the existence of nonprovable statements in certain types of formal logic. Study this from the point of view of how various mathematical concepts appear in the argument.

3.31. What analytic and algebraic concepts are applied to geometry in Euclidean, affine or projective geometry, algebraic geometry, combinatorial topology, and differential geometry? What geometric concepts are relevant to the implicit function theorem, the theorem on functional dependence?

3.32. Certain concepts of circuit theory, especially those associated with the time behavior of electrical quantities, have been associated with the Laplace transform. Describe them. What other mathematical concepts are used in circuit theory?

3.33. Diagram pure mathematics in a form in which independent axiomatization corresponds to a process and the topics are the results. Associate geometry and the various scientific theories with this diagram.

3.34. What problems were considered by Archimedes? Cavalieri? Wallis? Newton? Euler? Laplace? Gauss? Cauchy? Riemann? Weierstrass? Cantor? Borel? Lebesgue? Hilbert? L. Schwartz?

3.35. Consider any elementary development of group theory. Should elementary arithmetic be included in the axiomatic basis?

3.36. Beginning with telegraphy, describe the relationship of scientific understanding and communication.

References

1. Church, Alonzo, *Introduction to Mathematical Logic*, Princeton University Press, Princeton, New Jersey (1956).
2. De Wulf, Maurice, *The System of Thomas Aquinas* (reprint), Dover Publications, Inc., New York (1959).
3. Heath, T. L., *The Thirteen Books of Euclid's Elements* (reprint), Dover Publications, Inc., New York (1956).
4. Joad, C. E. M., *Guide to Philosophy* (reprint), Dover Publications, Inc., *New York* (1957).
5. Landau, Edmund, *Foundations of Analysis* (English translation by F. Steinhardt), Chelsea Publishing Company, New York (1951).
6. Morrow, G. R., *Proclus: A Commentary on the First Book of Euclid's Elements*, Princeton University Press, Princeton, New Jersey (1970).
7. Russell, Bertrand, in: *The Growth of Mathematics* (Robert W. Mark, ed.), Bantam Books, New York (1964).
8. Styazhkin, N. I., *History of Mathematical Logic from Leibnitz to Peano*, The M.I.T. Press, Cambridge, Massachusetts (1969).
9. Takeuti, G., and Zaring, W. M., *Introduction to Axiomatic Set Theory*, Springer-Verlag, New York (1971).
10. Whitehead, A. N., and Russell, B., *Principia Mathematica*, Cambridge University Press, Cambridge, England (2nd Edition, 1929; paperback partial reprint, 1962).

4

Ancient Mathematics

4.1. Ancient Arithmetic

We are interested in the improvement of mathematical understanding. Mathematical understanding is a necessary support for most complex cultures and is usually incorporated into both basic and technical education. Thus our present mathematical education has a layer structure, with the lower layers corresponding to the most widespread needs. In general the mathematical education of a culture is an indicator of its technical aspects. Since educational material tends to survive because there is so much of it, it is an excellent basis for the study of the growth of mathematical understanding.

There is a superficial correspondence between the historical development and the present layers of mathematical education, but this is misleading unless checked against the historical record. The historical record is much more revealing of the growth of understanding that was due to the activities of mature people in definitely nonscholastic environments. Mathematics itself has been continuously readjusted even in its most elementary aspects so that the main emphasis in the study of the historical development must deal not with the addition of layers but with the many metamorphoses of the conceptual structure. Our elementary arithmetic in its present form does not predate 1650 A.D., and yet it has had an evolutionary development that certainly began before the dawn of history.

The initial basis of arithmetic is counting. This capability certainly preceded the keeping of written records. The effective use of counting normally requires the aggregation of unit objects into larger units. Such aggregation is indicated in the names of numbers like twenty, thirty, or forty, and aggregation must have been a part of counting this high. Language also contains many examples of aggregation such as dozens, stones, scores, etc. that serve practical purposes. The elementary operations of addition and multiplication are further consequences of aggregation and are usually associated with a further step in which one has systematic and repeated aggregation, for example, the aggregation into tens, hundreds, thousands, etc.

The earliest written records dealing with arithmetic are from Egypt and Babylonia. There are a number of manuscripts from Egypt, the most important of which is the Rhind papyrus. It was written before 1700 B.C. by a priest, Ah'mose, and was presumably used for instructional purposes since much of it is in the second person. This document is translated and discussed in *The Rhind Mathematical Papyrus* by Chace *et al.*[2]

Another ancient papyrus, referred to as the Moscow papyrus, is discussed in "Mathematische Papyrus des staatlichen Museums der schönen Künste" by Struve.[11] Ancient Egyptian arithmetic is discussed in "Mathematics in Ancient Egypt" by Peet[10] and in *Science Awakening* by Van der Waerden.[12] Mathematical tablets from Babylonia are treated in *Mathematical Cuneiform Texts* by Neugebauer and Satz.[9]

In the Rhind papyrus, natural numbers are expressed in a decimal system, which is not a place system but one in which different symbols are used for 1, 10, 100, etc. For example, one could use | to correspond to one, ∩ to ten, and **C** to a hundred, and 241 would be written | ∩∩∩∩**CC**. The rule for addition in this notation is obvious. However, the calculation with natural numbers is more or less assumed in this document and numbers used are "mixed numbers," that is, natural numbers plus a fraction. This fraction is expressed as a sum of terms in the form $2/3$ or $1/n$, $n=2, 3, \ldots$. Obviously, the conceptual basis for this form of expression is quite different from that for our usual numerator–denominator fractions but precisely what it is is not clear, except that a process of choosing a smaller unit is involved.

Multiplication of natural numbers was based on a process of repeated doubling, i.e., of adding a number to itself. Suppose one wishes to multiply 231 by 19. One doubles the multiplicand repeatedly, writing the result beside the corresponding multiplier:

1	**231**
2	**462**
4	924
8	1848
16	**3696**
19	4389

One stops with the highest multiplier that does not exceed the multiplier of the original problem. One then checks the multipliers that add up to the original multiplier (indicated in boldface above) and adds the corresponding multiples of the multiplicand. This yields the product, 4389.

The division procedure is similar. Suppose one wishes to divide 2783 by 39. The procedure is based on doubling the divisor

1	39
2	78
4	**156**
8	312
16	624
32	1248
64	**2496**
71	2769

until the next double would exceed the given dividend. One then adds a selection of these multiples until the sum is less than the dividend by a quantity less than 39. This sum, 2769, corresponds to 71×39 with a remainder of 14. One can continue the multiplication of 39 with fractions:

$\frac{1}{3}$	13
$\frac{1}{39}$	1

$$71 + \tfrac{1}{3} + \tfrac{1}{39} \qquad 2783$$

This Egyptian procedure was utilized for many centuries.

In modern terms, this procedure involves expressing the multiplier in binary form, but it would perhaps be more appropriate to say that the simplest form of multiplication is to double a quantity.

4.2. Egyptian Mathematics

In Ah'mose, the operations on the integers are assumed to be available. The main concern is with operations on mixed quantities. For these a number of problems are apparent in order to obtain an effective arithmetic equivalent to that for the rational numbers. Quite a number of applications are discussed in the Rhind papyrus and the author is clearly interested in the procedures as much as in the problems themselves.

One problem is to express $2/n$, where n is an odd number, as a sum of fractions with numerator 1 (this is required for addition and for multiplication since multiplication is based on adding a number to itself). One can always solve this problem since

$$\frac{2}{2k+1}=\frac{1}{k+1}+\frac{1}{(k+1)(2k+1)},$$

but in general this is not the preferred solution, since the denominator in the last term may be large. The Rhind papyrus begins with a discussion that obtains the appropriate expression for $2/n$ for $n=5, 7,..., 101$, and different procedures are used in various cases so that one has the equivalent of a mathematical investigation in which the terms $1/n$ are clearly entities. Other problems are to add a number of such expressions when the sum exceeds 1 or to subtract such an expression from 1. The procedure usually involves choosing one of the denominators and using the corresponding fraction as a new unit in terms of which the other fractions are expressed, possibly as mixed numbers. For example,

$$\frac{1}{2}+\frac{1}{3}+\frac{1}{8}+\frac{1}{9}=\frac{1}{9}\left(4\tfrac{1}{2}+3+1\tfrac{1}{8}+1\right)=\frac{1}{9}\left(9+\frac{1}{2}+\frac{1}{8}\right)$$

$$=1+\frac{1}{18}+\frac{1}{72}.$$

The basis of the Egyptian economy was the production of grain, which was consumed either as bread or beer. The equivalence between units of these commodities is based on the amount of grain in them, and this leads to a number of problems in proportion. Plans were made on an annual basis, e.g., the amount of grain needed for a certain number of people was estimated. Another type of problem involved a division of supplies on an unequal basis, which leads to a problem in arithmetical progression. The relative values of precious metals produces problems of exchange. If one wishes to anticipate the amount of grain available from a given field, one must estimate the area. There is also the matter of taxes. The storage of grain raises questions concerning the volume of a corn crib whose dimensions are given.

The problems associated with addition are indicated above. Addition will yield multiplication, but division (or its equivalent) requires further consideration. For example, suppose one is asked to find a quantity such that itself and one-seventh of itself total 19. The number, $1+\frac{1}{7}$, is considered to be the operation of multiplying by this quantity. When applied to 7, it yields 8. Now 8 is contained $2+\frac{1}{4}+\frac{1}{8}$ times in 19. Thus, the unknown quantity must contain 7 the same number of times, i.e., the answer is $16+\frac{1}{2}+\frac{1}{8}$. The number $1+\frac{1}{7}$, therefore, is handled as a proportion. Other examples given in the papyrus correspond to division by $1+\frac{2}{3}+\frac{1}{2}+\frac{1}{7}$, $1+\frac{1}{3}+\frac{1}{4}$, and $3+\frac{1}{3}$.

Rectilinear areas are readily handled in the rectangular case. Circular area and cylindrical volume are effectively handled from the practical point of view. If d is the diameter, the area of a circle is taken to be $(\frac{8}{9}d)^2$. There are also problems in proportion associated with the various measurements of a pyramid. Certain problems deal with numbers as such, but others are expressed in terms of measures of grain, loaves of bread, measures of beer, etc. We have, therefore, an arithmetic, but the association with the practical situation is close.

The obvious purpose of the manuscript is to describe procedures for solving problems of various types. The elements of the problem are given numerical values, and operations on these are described in terms of these values. The directions for the procedure are given in the second person. You are told to do this or that with these numbers. For example, the student is told that when he has reached the exalted estate of scribe and someone requests the area of a circular field that is 9 cubits across, then he is to take $\frac{1}{9}$ of the 9 cubits, which is 1, and subtract it from the 9 cubits and square the result. Lest the feeble-minded student confuse the 9 cubits with the $\frac{1}{9}$ that appears, the problem is repeated with a diameter of 10 cubits.

Thus, the notion of an algorithm is clear and also the notion of an arithmetic procedure separated conceptually from the problem. There is, of course, no formulation in algebraic symbols, no symbol for an "arbitrary number," and no symbolic method for handling this concept.

4.3. Babylonian Mathematics

Babylonian mathematics differs from the Egyptian in a number of aspects and probably should be considered as more sophisticated. Their arithmetic also involved mixed numbers, but they are expressed in the sexagesimal system, and the fractional part corresponds to a fraction with a denominator in the form $2^p3^q5^r$. The sexagesimal system was a place system in which the numbers in each place were expressed decimally. Thus, one needed only addition and multiplication tables for the digits, 1, ..., 9, and for 10, 20, ..., 50. But there were ambiguities insofar as there was no zero digit or sexagesimal point, and these ambiguities had to be resolved from the context.

Thus, the Babylonian arithmetic was quite different from the Egyptian described above. Procedures were much more formally organized around tables. For the purposes of division, reciprocals were used. The most elementary tables would list the "regular numbers," i.e., those in the form $2^p3^q5^r$ and their reciprocals. These tables would therefore not contain reciprocals for 7, 11, 13, etc. But there were also more advanced tables containing approximations to the reciprocals of numbers with these primes as factors and also reciprocal tables obtained by repeated doubling and halving from a given pair. Multiplication tables contain the multiples $2c, 3c, ..., 19c, 20c, 30c, 40c,$

$50c$ for given numbers c. This is not the minimum necessary for multiplication, but it was probably more convenient. The article by Neugebauer and Satz[9] describes the available tablets and the form assumed by these multiplication tables, some of which are quite extensive. There are also tables for squares, square roots, and cube roots, and these are highly significant for the kind of problems dealt with. There are also tables of powers.

Geometric problems are associated either with areas, as in surveying problems, or with volumes, as in problems concerning the excavation of canals and irrigation ditches. The effort needed is estimated in terms of man-days and the cost in the form of the corresponding wages to be paid in silver. The excavation volume is divided into three layers, and the work for a given volume depends on the depth of the layer. Other volume problems involve the number of bricks in a pile of given volume. Problems tend to have an economic aspect, which one would expect if they are the concern of administrators. Area problems of nonrectangular areas are handled only approximately.

Problems are presented in concrete terms, and algorithms are indicated in terms of operations on specific numbers when the procedure for solving problems is explicitly given. The solution procedure is in the second person. There is a wide range of problems in which one seeks for two unknown quantities—"length" x and "width" y—for which the "area" xy is given and a linear combination described by a sequence of numerical operations is given. We would deal with such a problem by solving the linear combination for one variable, substituting in the other equation, and solving the quadratic equation. The solutions given follow precisely the formula for solving the resulting quadratic equation, but there is no hint as to the intermediate process. It has been suspected that the relation $[(x+y)/2]^2 = xy + [(x-y)/2]^2$ was inferred on geometrical grounds, possibly in the form $(a+b)^2 = 4ab + (a-b)^2$. Thus in Figure 4.1, this relation becomes immediately obvious if one moves the area I to the position II. But the range of problems requires some more sophisticated procedures. Notice that the numerical procedure corresponding to the formula for solving quadratic equations requires square roots, and, of course, tables for both the square and square roots were used. There are also straight linear problems and certain rather ubiquitous arithmetic series problems.

A very interesting table is one that gives a sequence of triplets of numbers that satisfy $l^2 + b^2 = d^2$ arranged so that d^2/l^2 is increasing. Apparently these triplets were obtained by means of the formulas $l = 2pq, b = p^2 - q^2, d = p^2 + q^2$, where p and q are "regular" numbers, that is, have precise reciprocals in the sexagesimal system. There is no indication as to the purpose of this table, but a modern table of trigonometrical functions is also a table of right triangles. The ancient table is a table of right triangles arranged according to the secant.

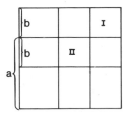

Figure 4.1. Geometric algebra construction.

The sexagesimal system of Babylonia has survived in our measurements of angles and time. There are historians who believe that the geometric algebra of the Greeks developed from Babylonian procedures. Astronomy can be readily traced back to Babylonia. The Greeks claimed Egypt and Babylonia as ancient sources for their mathematics. The methods of the latter are certainly more than primitive, but the step to the classical mathematics of the Greeks was enormous.

For further discussion of the material in this section, see Neugebauer and Satz.[9]

4.4. Greece

The historical development of ancient civilizations is a fascinating complex interaction of many elements, including trade, wars, and inventions. One of the most important of the inventions was that of the alphabet by the Phoenicians. This was the core element that integrated the culture of the Greek cities of the Mediterranean.

We must limit ourselves not only to mathematics but even to certain aspects of mathematics, in particular its conceptual character. One classical development was that of centers of higher education. Formal education for the well-to-do was by tutors in its elementary stages. But for those who wanted to go further there was travel and certain centers of learning where one could study one or more subjects, such as rhetoric, law, philosophy, and, most universally, mathematics.

In Athens around 400 B.C., Plato founded the Academy, which continued until 529 A.D., when it was closed by an edict of Emperor Justinian. Although the Academy did not have the organization of a modern university, it had formal lectures, produced handwritten manuscripts, held property, and had a rather respectable endowment. Aristotle founded a rival institution called the Lyceum. In Alexandria there was the famous Museum, or Temple of the Muses, with a library and staff of scholars supported for many centuries by the Ptolemies and later by the Roman emperors. The mathematics of classical times is associated with Alexandria in one or more ways.

Mathematics played an introductory role in the classical higher educational program. The name itself is from the word for learning, $\mu\alpha\theta\eta\sigma\iota\sigma$, and comprised geometry, arithmetic, mechanics, astronomy, optics, geodesy, music, and calculation. According to the Greeks, geometry was first discovered by the Egyptians and was used by them to resurvey their land after the Nile inundations. The Phoenicians are associated with number theory because of the necessities of trade and exchange.

Men like Thales of the city of Miletes (625–545 B.C.), who traveled to Egypt and Mesoptamia and studied there for a number of years, were credited with introducing mathematics to the Greeks. The names of various early mathematicians are known from later references, but the mathematics of this initial period is known to us only in the form of Euclid's *Elements*. Euclid (330–250 B.C.) lived in Alexandria in the time of the first Ptolemy. His *Elements* apparently played the role of a basic textbook and also a treatise of fundamental results on which the more sophisticated works of Archimedes and Apollonius were based.

An appreciation of classical mathematics is best obtained from commentaries directly associated with translations of the original works when they are available. Probably one should begin with *The Thirteen Books of Euclid's Elements* by Heath.[5] We have mentioned Van der Waerden's[12] book, which discusses the attainments of Greek mathematics in detail and with their procedures of proof. Much of our information concerning ancient mathematics is in the commentary of Proclus, who also represents a Platonic viewpoint many centuries after Plato; see *Proclus: A Commentary on the First Book of Euclid's Elements* by Morrow.[7] Archimedes' work is discussed in *Archimedes* by Dijksterhuis.[4] The successes and limitations of Greek mathematics from a modern point of view are discussed in *The Role of Mathematics in the Rise of Science* by Bochner.[1]

4.5. Euclid's Elements

Let us now consider the various aspects of Euclid's *Elements* that do not appear in the available versions of previous mathematics. First, there is the logical structure. There is a step-by-step development beginning with "first principles" and yielding a sequence of proven statements. Euclid's *Elements* presents an abstraction from spatial experience, termed "geometry," which involves two new characteristics. One of these is constructive and contains many results of great practical importance. But geometry deals with magnitudes, and magnitudes require a subtle extension of the basic conceptual complex of objects, sets, and natural numbers. The notion of relative size of two magnitudes of the same kind may not be expressible as a ratio of natural numbers.

From Euclid on, the logical aspect has been the distinguishing characteristic of mathematics and geometry, the bridge between elementary and "higher" education. The "first principles" are presented in three forms— definitions, postulates, and "common notions" or axioms. The main burden of expressing the conceptual structure is placed on the definitions. These occur at the beginning of every book, except Books VIII, IX, XII, and XIII of the *Elements*. In many cases they are not simply definitions in the modern sense of abbreviations or names for previously developed concepts. Instead they are efforts to describe the conceptual structure on which the discussion is based. They formulate the abstraction process by which spatial experience is condensed into certain essential elements and is expressed by diagrams.

Thus, we have the definitions for a point (I.1. A point is that which has no part), for a line (I.2. A line is a length without breadth), for a surface (1.5. A surface is that which has length and breadth only), and a straight line (I.4. A straight line is a line which lies evenly with the points on itself). Somewhat similar is the definition of an angle (I.8. A plane angle is the inclination to one another of two lines in a plane which meet each other and do not lie in a straight line); this definition is also specialized into the notion of rectilinear angle, i.e., one whose sides are straight lines. Rectilinear angles in Euclid are treated as magnitudes. In modern geometric treatments, these notions would be considered "undefined terms" with postulated relations between them.

Other definitions explain relationships (I.3. "The extremities of a line are points"). There is, of course, a large number of definitions describing plane figures and spatial configurations, for example, right angle, acute angle, obtuse angle, plane figure, circle, diameter, center, semicircle, triangle, and quadrilateral.

The constructive geometric development is based on definitions and on the postulates. There are five postulates. Postulates 1, 2, and 3 state that one can construct a line segment with given end point, a straight line through two points, and a circle with a given center and radius. Postulate 4 states that all right angles are equal. Postulate 5 is, "If a straight line falling on two straight lines, makes the interior angles on the same side less than two right angles, the two straight lines, if produced indefinitely, meet on that side on which are the angles less than two right angles."

At first glance, Postulate 4 seems to be similar to the proven statements, and since the logical discussion needs a starting point, this would seem to be its role. Postulate 5 appears to be another "constructive" postulate, i.e., it permits one to construct a point. However, the situation is more subtle than this and we will return to it.

The geometry based on these definitions and postulates is a sequence of propositions. A discussion of mathematical reasoning and the logical structure of propositions is given in Proclus (see Morrow[7] pp. 159ff.; the numbers in the margin refer to the Friedlein text and references to Proclus in Heath[5] also follow this text). In F. 206 he tries to fit the reasoning to the syllogistic form and also "cause and effect" in the sense of "essential cause." But in F. 207 he plunges into an entirely different explanation.

> Furthermore, mathematicians are accustomed to draw what is in a way a double conclusion. For when they have shown something to be true of a given figure, they infer that it is true in general going from the particular to the general conclusion. Because they do not make use of the particular qualities of the subject but draw the angle or straight line in order to place what is given before our eyes, they consider what they infer about the given angle or straight line can be identically asserted for every similar case. They pass therefore to the universal conclusion in order that we may not suppose that the result is confined to the particular instance. This procedure is justified, since for the demonstration, they use the objects set out in the diagram not as these particular figures but as figures resembling others of the same sort. It is not having such-and-such a size that the angle before me is bisected, but as being rectilinear and nothing more. Its particular size is a character of the given angle, but its having rectilinear sides is a common feature of all rectilinear angles. Suppose the given angle is a right angle. If I used rightness for my demonstration, I should not be able to infer anything about the whole class of rectilinear angles; but if I make no use of its rightness and consider only its rectilinear character, the proposition will apply equally to all angles with rectilinear sides.

A complete proposition according to Proclus (F. 203) should contain "an enunciation, an exposition, a specification, a construction, a proof and a conclusion," although the most essential parts are the enunciation, proof, and conclusion. The "enunciation" is, of course, the statement of the proposition.

> The exposition takes separately what is given and prepares it in advance for use in the investigation. The specification takes separately the thing that is sought and makes clear precisely what it is. The construction adds what is lacking in the given for finding what is sought. The proof draws the proposed inference by reasoning scientifically from the propositions that have been admitted. The conclusion reverts to the enunciation, confirming what has been proven.

Thus, geometrical reasoning is an abstracted intellectual experience in which the diagram provides both a symbolic record and support for the imaginative development. The general statements are given a specific diagram interpretation in the exposition and specification. Our intellectual experience is with the specific diagram. Further elements of the diagram are then introduced in the construction, which must be based on the postulates (F.209. In general, the postulates contribute to the construction and the axioms to the proofs). The proof then follows by considering the properties of the completed figure as known either from the axioms, the postulated properties of the constructed elements, or previously proven propositions. The choice of the constructed figure may represent considerable ingenuity and analysis of the geometrical relations.

The constructive geometry of Euclid included the theory of congruent figures and the construction of the regular polygons of three, four, five, and six sides and the five regular polyhedra. We have mentioned the geometric algebra of the Sumerians in which algebraic problems are formulated in geometric terms. The corresponding procedures are, of course, available in Euclid. But now Euclid can present constructive versions of the algorithms and prove the desired relations.

4.6. Magnitudes

In our experience, it is normally necessary to deal with the notion of amount. We buy some things by length, others by area, volume, or by weight. We usually handle this situation by analogy with the situation in which we deal with discrete objects. We choose a "unit amount" and consider that we are dealing with an integral multiple of this amount. But this choice of "unit amount" is clearly arbitrary and our quantities in general come in "odd lengths." While this procedure may be a practical simplification, it does not correspond to a satisfactory formulation of the experience pattern.

Thus, we must deal with the concept of a magnitude and a precise discussion must represent an experience pattern that extends the elementary logical combination of objects, sets, and natural numbers. The axiomatic approach of Euclid does permit one to consider the notion of magnitude in a satisfactory way. We assume that we can compare two magnitudes of the same kind to see which is the larger and determine when one is an integral multiple of the other and the inverse relationship, i.e., an "aliquot part." Essentially, Euclid considers the full range of experience with these processes to yield a notion of relative size or ratio of two magnitudes of the same kind.

The magnitudes that appear in Euclid's elements include lengths, angles, areas, and volumes. Comparison in size is possible between such magnitudes, and forming multiples is also possible. However, it had also been shown that there exist pairs of magnitudes that are not multiples of a common unit, for example, the side and diagonal of a square. Thus, the arithmetic of rational numbers is not adequate.

The discussion of magnitudes is based on certain definitions and on the "common notions," or axioms. The "common notions" describe properties of equality. "Equality" in Euclid has a number of distinct meanings, and these axiomatic properties apply to all of them. Thus, equality refers to identity, congruence, equality of magnitudes (as when a triangle is said to equal a parallelogram because they have equal area), and equality of the ratios of magnitudes. The first five axioms are

1. Things equal to the same thing are equal to each other.
2. If equals are added to equals, the wholes are equal.
3. If equals are subtracted from equals, the remainders are equals.
4. Things that coincide with each other, are equal to each other.
5. The whole is greater than the part.

In Book V, Definition 3, a ratio is defined as a "relation in respect to size between two magnitudes of the same kind." The notion of integral multiple is defined in Definition 2. The equality of two ratios is specified by Definition 5. In modern symbols $a : b = c : d$ if for every pair of integers m and n the multiples ma and nb are related as either $ma < nb$, $ma = nb$, or $ma > nb$ accordingly as $mc < nd$, $mc = nd$, or $mc > nd$. In Definition 7, $a : b$ is said to be greater than $c : d$ if there is a pair of integers m, n such that $ma > nb$ and $mc \leqslant nd$. The effect then of the definition of equality is to consider for each ratio a division of the pairs of integers (i.e., the fractions) into three disjoint sets and define equality as corresponding to the same division of the set of fractions. This, of course, corresponds to the modern definition of real numbers as essentially just this same division of the fractions into subsets.

In classical geometry, a ratio is a relationship not an entity. Later mathematicians were to introduce the notion of a real number as a ratio of a special kind of magnitude, i.e., line segments on a straight line. Consider the Euclidean straight line; specify a point O as origin and a line segment OP_1 as unit. Then for every point P on the line there is the "real number x" as the ratio of OP to OP_1 (this included a notion of sign as a later development). One can regard as intuitive that every ratio has a ratio of this type equal to it, i.e., they specify the same division of the fractions into sets. Thus, given a magnitude of any kind and the appropriate unit of the same kind, we can associate a size as the real number corresponding to the ratio of the magnitude to the unit.

It seems natural, therefore, to find in Euclid a considerable discussion of the properties of natural numbers and what we now consider to be the algebra of ratios. Thus, most of the results in number theory that precede the unique factorization theorem appear and are given proofs in terms of multiples of a line segment. The basic result is the Euclidean algorithm for finding the greatest common divisor of two numbers. "Number theory" permits Euclid to show the irrationality of certain ratios and develop a theory of quadratic irrationalities.

The notion of magnitude and ratio permits one to deal with the possibility that the ratio of the circumference of a circle to its diameter is not rational and to approximate it by geometrical constructions as closely as desired. One can inscribe and circumscribe polygons relative to a circle for this purpose; this corresponds to squeezing down on the desired ratio as tightly as desired. Relations were obtained in classical times equivalent to our present formulas for the areas of triangles, regular polygons, circles, cones, spheres, and zones on spheres and the volumes of prisms, tetrahedra, cones, frustums of cones, and spheres.

Thus, the total formulation found in Euclid with its range of geometric information is a stunning intellectual achievement. The range of experience patterns presented was a cultural inheritance basic to all subsequent civilization. Not the least important of these was the notion of an "axiomatic development" that would certainly permit independent development, but science was to advance by incorporating the Euclidean structure in a larger framework of "natural philosophy." The arithmetic of aggregation and accounting still had an independent existence, but it was now linked to a much more powerful partner.

4.7. Geometry and Philosophy

The Grecian geometer considered his geometry as a true description of space, and his arguments are efforts of his mind to reach the truth. Aristotle considered the abstraction process as dealing with forms associated with the essential nature of the object. Plato considered the forms as representing a higher ideal truth than that associated with actual mundane objects. There is a considerable discussion of Aristotle's views in Heath.[5] The classical notion of abstraction is discussed in De Wulf's[3] *The System of Thomas Aquinas*. The views of Plato on mathematics are of considerable importance and we quote from the translation of B. Jowett[6] from the end of Book VI of the *Republic*:

>...You are aware that students of geometry, arithmetic, and the kindred sciences assume the odd and the even and the figures and three kinds of angles and the like in their several branches of science; these are their hypotheses, which they and everybody are supposed to know, and therefore they do not deign to give any account of them either to themselves or to others; but they begin with them, and go on until they arrive at last, and in a consistent manner, at their conclusion....
>
>...although they make use of visible forms and reason about them, they are thinking not of these, but of the ideals which they resemble; not of the figures which they draw, but of the absolute square, and the absolute diameter, and so on—the forms which they draw or make, ... are converted by them into images but they are really seeking to behold the things themselves which can only be seen with the eye of the mind?...

And of this I spoke as the intelligible, although in the search after it the soul is compelled to use hypotheses; not ascending to a first principle, because she is unable to rise above the region of hypothesis, but employing the objects of which the shadows below are resemblances in their turn as images, they having in relation to the shadows and reflection of them a greater distinction and therefore a higher value....When I speak of the other division of the intelligible, you will understand me to speak of that other sort of knowledge which reason itself attains by the power of the dialectic, using the hypotheses not as first principle but only as hypotheses—that is to say as steps and points of departure into a world which is above hypotheses, in order that she may soar beyond them to the first principle of the whole; and clinging to this and then to that which depends on this by successive steps she descends again without the aid of any sensible object from ideas, through ideas and in ideas she ends.

In Book VII of the *Republic* the educational values of geometry from the intellectual point of view are stressed, and there is a minor concession relative to its practical value.

It must be admitted that the logical ideal of inferences based purely on previously established results is not realized in Euclid, and these discrepancies are considered carefully in Heath. Questions of interior and exterior of a figure and indeed certain intersection properties are assumed from diagrammatic experience. Heath also discusses the relationship of Euclid's "definitions" with the norms of the logic of Aristotle. In modern geometry, these gray areas of Euclid have been replaced in many different ways by precise axiomatic treatments of "incidence geometries" or geometries involving analytic notions such as the cross product.

The choice of the initial principles of geometry was clearly based on patterns of experience, and it was reasonable to assume that this abstraction process was a method of grasping the truth. But the use of *a priori* knowledge to base a theory was subject to a number of weaknesses, irrespective of how firmly this knowledge was based on our intuition, which presumably is an integration of past experience. The information used was selected under various pressures to make the resultant theory conform to preset philosophical, religious, economic, or political theories. This objection does not seem to be applicable to geometry, but in geometry the fifth postulate was an absolute relationship whose verification would require unattainable accuracy. It is equivalent to the statement that the sum of the angles of a triangle is a straight angle. Thus, it is impossible to be sure by measurement that one has this rather than a slightly different case, which would correspond to a non-Euclidean geometry.

Thus, principles based on previously available knowledge require experimental verification, and this combination corresponds to an inductive process for establishing the postulates of an acceptable theory. But both the range of available experience and the precision of the experimental verification are limited at any time, so that while a theoretic advance may produce a considerable expansion of understanding at a certain time, our experience is that new limits will appear and we deal with repeated adjustments. Our experience with space and time has had this character.

4.8. The Conic Sections

Classical geometry is fascinating in its own right. One can obtain an excellent introduction from the works of Heath,[5] Van der Waerden,[12] and others, and one can use translations to go deeper into the works of Archimedes and Apollonius. The *Commentary* of Proclus[7] is excellent reading. We will briefly discuss the conic sections, since the results on them had very important later effects.

For us the geometric relations involved are most readily handled by means of the modern algebraic equivalents. One effective way of expressing geometric relations is in terms of perpendicular line segments. For example, a circle can be described by a relation

$$y^2 = x(2r - x),$$

where $(x, 2r - x)$ is a division of a diameter and y a perpendicular half chord at the division point. Such a relation was called a "symptom." For an ellipse one has the symptom

$$y^2 = \alpha x(2a - x)$$

for some value of α with $0 < \alpha < 1$. Similarly the hyperbola has a symptom

$$y^2 = \alpha x(2a + x)$$

and the parabola

$$y = 2px.$$

Clearly this concept is similar to our notion of the "equation of the locus" in Cartesian coordinates (see Van der Waerden,[12] pp. 241ff).

Originally the conic sections were obtained by intersecting right circular cones by planes. A right circular cone is a cone such that a plane perpendicular to the axis intersects the cone in a circle (or at the vertex). A plane that intersects one nappe only of the cone will yield an ellipse, except when it is parallel to an element, in which case the intersection is a parabola. If the plane intersects both nappes, the intersection is a hyperbola. The symptom is readily obtained from this definition.

Apollonius of Perga showed that the plane section of any circular cone was a plane section of a right circular cone, i.e., a conic section in the restricted sense. To obtain this more general result, one must generalize the notion of a symptom. For the symptom in the previous case, x is laid off on the major axis and the direction for y is taken perpendicular to this axis. There is a corresponding relation in which one takes x along any central diameter and measures y in a fixed direction from P to the locus, but this fixed direction in general will not be perpendicular to the chosen diameter.

For the ellipse, one can readily find the pairs of directions that yield such a symptom. Let us consider the standard symptom on the major axis, $y^2 = \alpha x(2a - x)$. It is convenient to replace x by $x' = a - x$ so that one has the central form $y^2 + \alpha x'^2 = \alpha a^2$. We drop the prime. Now consider any diameter, i.e., line through $(0, 0)$. This will intersect the ellipse in the points (x_1, y_1), $(-x_1, -y_1)$. Let (x_2, y_2) be chosen on the ellipse so that $y_1 y_2 + \alpha x_1 x_2 = 0$. It is an interesting exercise to show that this can always be done.

If we express an arbitrary point (x, y) vectorially in terms of (x_1, y_1) and (x_2, y_2), we have $(x, y) = s(x_1, y_1) + t(x_2, y_2)$ or

$$x = sx_1 + tx_2$$

$$y = sy_1 + ty_2.$$

The condition, $y^2 + \alpha x^2 = \alpha a^2$, that (x, y) be on the ellipse, becomes

$$\alpha a^2(s^2 + t^2 - 1) + 2st(y_1 y_2 + \alpha x_1 x_2) = 0.$$

Thus, the condition becomes $s^2 + t^2 = 1$.

If $P_1(x_1, y_1)$ is on the ellipse, then the points $s(x_1, y_1)$ for $-1 < s < 1$ are on the chord joining P to $(-x_1, -y_1)$. For each such value of s there are two points $(sx_1, sy_1) \pm (1 - s^2)^{1/2}(x_2, y_2)$ on the ellipse. The line joining these two points is parallel to the chord l_2 joining (x_2, y_2) and $(-x_2, -y_2)$. If $s^2 > 1$ there is no ellipse point on the line through (sx_1, sy_1) parallel to l_2. For $s = \pm 1$ there is just one such point on the corresponding line parallel to l_2. These two latter lines are obviously limiting positions of a secant as it moves parallel to itself and thus are tangents. Thus, these elementary algebraic procedures, which involve at most quadratic relations, permit one to obtain tangents.

For the parabola, $y^2 = 2px$, the alternate diameters that may be used are all parallel to the original axis. Let (x_0, y_0) be a point on the parabola other than $(0, 0)$ and let

$$(x, y) = (x_0, y_0) + s(2x_0, y_0) + t(1, 0)$$

or

$$x = (1 + 2s)x_0 + t, \quad y = (1 + s)y_0.$$

If we fix t and let s vary, we get a line, through $(x_0 + t, y_0)$ with slope $y_0/2x_0$. The condition that (x, y) be on the parabola reduces to $s^2 x_0 = t$, and hence, for $t > 0$ there are two points on this line. For $t = 0$ we obtain the result that the corresponding line is tangent to the parabola at (x_0, y_0).

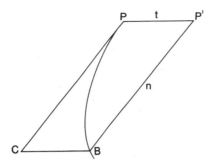

Figure 4.2. Alternate diameters for parabola.

The relation $s^2 x_0 = t$ can also be expressed in terms of lengths. Let n be the length of $s(2x_0, y_0)$; i.e., $n^2 = s^2(4x_0^2 + y_0^2) = s^2(4x_0^2 + 2px_0)$. Then $s^2 x_0 = t$ becomes $n^2 = (4x_0 + 2p)t$. We can make the following geometrical interpretation (see Figure 4.2). Let P be a point on the parabola and let C be a point on the tangent at P. Let B be the intersection of the parabola with the line through C parallel to the axis. If we complete the parallelogram $PCBP'$, we have $PC = P'B = n$ and $CB = PP' = t$. Thus, $(PC)^2 = (4x_0 + 2p)BC = lBC$, where l does not depend on C.

In the above, we have permitted algebraic convenience to divert us from the precise analogs of the geometric discussions. But one can point out that the ancient geometers had a facility with their type of discussion that certainly was as good as that associated with our algebra, and we can only appreciate the fascination and interest of this geometry when we can handle these problems with some ease.

For further discussion of the material in this section, see Dijksterhuis,[4] Heath,[5] Morrow,[7] and Van der Waerden.[12]

4.9. Parabolic Areas

The concept of a tangent, of course, was highly significant for the future, although, of course, new methods not tied so tightly to a geometric interpretation would be required to yield the differential calculus. On the other hand the integral calculus arose directly from Archimedes' procedures and the original arguments are highly significant. The initial problem considered was determining the area bounded by the arc of a parabola and a chord. We show first the following:

Lemma. (See Figure 4.3.) Let $P_1 P_2$ be a chord of a parabola. Let $P_2 Q$ be tangent to the parabola, and let QE and CG be parallel to the axis of the

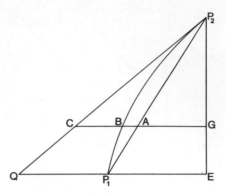

Figure 4.3. Archimedes' lemma.

parabola. Let P_2E be perpendicular to QE. Then

$$\frac{AB}{AC}=\frac{GE}{EP_2}.$$

Proof.   From the argument following the discussion of conjugate axes for the parabola, we obtain an l such that

$$(P_2C)^2=lCB,\qquad (P_2Q)^2=lQP_1.$$

Thus,

$$\frac{CB}{QP_1}=\left(\frac{P_2C}{P_2Q}\right)\left(\frac{P_2C}{P_2Q}\right)=\frac{CA}{QP_1}\frac{P_2C}{P_2Q}.$$

Thus,

$$\frac{CB}{CA}=\frac{P_2C}{P_2Q}=\frac{P_2G}{P_2E},$$

which implies the equality of the Lemma.

One can now show that the parabolic area BP_2AP_1 is equal to one-third of the area of the triangle P_1P_2Q. Archimedes first gives a "heuristic argument." Let us take moments around QE. The above lemma yields $CA \cdot GE = AB \cdot EP_2$. Thus, the moment of CA around QE equals the moment obtained by placing AB at P_2. If we consider the areas as made up of parallel lines, then the moment of the triangle P_1P_2Q must equal that of the parabolic area P_1BP_2A placed at P_2E. Thus,

$$P_2E \cdot (P_1BP_2A)=\text{moment of } P_1P_2Q=P_1P_2Q \cdot \tfrac{1}{3}P_2E.$$

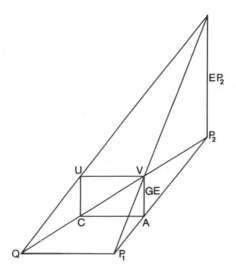

Figure 4.4. Moment tetrahedron.

The moment of P_1P_2Q around P_2Q can be shown to be equal to the volume of a tetrahedron with base P_1P_2Q with altitude equal to P_2E (see Figure 4.4).

Our modern procedure for finding the area BP_2AP_1 is by integration. Let $x = GE$. Then

$$CA = QP_1(P_2E - x)/P_2E$$

and from the lemma, $BA = QP_1(P_2E - x)x/P_2E^2$. The usual summation definition of the integral yields

$$(P_1BP_2A) = \int_0^{P_2E} BA\,dx = \tfrac{1}{6}QP_1 \cdot P_2E = \tfrac{1}{3}(P_1P_2Q).$$

Our modern interpretation of the integral is that of a limiting process. On the other hand the heuristic argument can be expressed in terms of "infinitesmals," that is, if we consider an integral as an infinite sum of "infinitesmals." Thus, the "lines" AC and AB are to be replaced by parallelograms with the given line as base and height dx. But Archimedes considered a heuristic argument such as the above as simply indicating the relation to be obtained, and a satisfactory proof is given by the method of "exhaustion."

Archimedes' procedure is described in Van der Waerden[12] (pp. 216ff). We present a variation of this argument. As in the integral approximation, we divide P_2E by points $G_0 = E$, G_1, ..., $G_n = P_2$. Corresponding to the point G_i we have points A_i, B_i, C_i. Let G_i' correspond to the midpoint of $G_{i-1}G_i$. Consider Figure 4.5 with KL tangent to the parabola at B_i'. We take $x_i = G_iE$,

$\Delta x = x_i - x_{i-1} = P_2E/n$. We can obtain by elementary procedures two formulas for the moment, M_i, of the quadrilateral $A_{i-1}C_{i-1}C_iA_i$:

$$M_i = \tfrac{1}{2}(C_{i-1}A_{i-1}x_{i-1} + C_iA_ix_i)\Delta x + \tfrac{1}{6}(C_{i-1}A_{i-1} - C_iA_i)\Delta x^2 \qquad (m_1)$$

$$= C_i'A_i'x_i'\Delta x - \tfrac{1}{12}(C_{i-1}A_{i-1} - C_iA_i)\Delta x^2. \qquad (m_2)$$

For example, M_i can be evaluated as the volume of an appropriate solid.

If we substitute in (m_1) the result from the lemma, i.e., $CAx = BAP_2E$, we obtain

$$M_i = \tfrac{1}{2}(B_{i-1}A_{i-1} + B_iA_i)\Delta x \, P_2E + \tfrac{1}{6}(C_{i-1}A_{i-1} - C_iA_i)\Delta x^2.$$

Now $\tfrac{1}{2}(B_{i-1}A_{i-1} + B_iA_i)\Delta x$ is the area of the quadrilateral $B_{i-1}A_{i-1}A_iB_i$ contained in the parabolic region with these vertices. Thus, if a_i denotes the area of this parabolic region, we obtain

$$M_i \leqslant a_iP_2E + \tfrac{1}{6}(C_{i-1}A_{i-1} - C_iA_i)\Delta x^2.$$

If we sum over i and let M denote the moment of P_1P_2Q around P_2E and a the area of the parabolic segment P_1AP_2B, we obtain

$$M \leqslant aP_2E + \tfrac{1}{6}P_1Q\Delta x^2.$$

Again, if we use (m_2) and the lemma, we obtain

$$M_i = B_i'A_i'\Delta x \, P_2E - \tfrac{1}{12}(C_{i-1}A_{i-1} - C_iA_i)\Delta x^2.$$

But $B_i'A_i'\Delta x$ is the area of the quadrilateral $KA_{i-1}A_iL$, which contains the parabolic region $B_{i-1}A_{i-1}A_iB_i$ and we have

$$M_i \geqslant a_iP_2E - \tfrac{1}{12}(C_{i-1}A_{i-1} - C_iA_i)\Delta x^2.$$

This yields

$$M \geqslant aP_2E - \tfrac{1}{12}P_1Q\Delta x^2.$$

Since n can be taken arbitrarily large, the assumption that either $M > aP_2E$ or $M < aP_2E$ leads to a contradiction. Provided one assumes the existence of M and a, the above argument is logically satisfactory. A form of the integral calculus developed from these procedures.

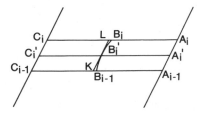

Figure 4.5. Slice of parabolic area.

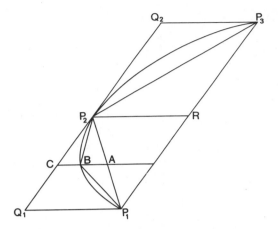

Figure 4.6. Archimedes' second proof.

Another argument of Archimedes for obtaining the area of a parabolic segment utilizes geometric series. Consider Figure 4.6: P_1 and P_3 are two points on the parabola; R is the midpoint of P_1P_3, and RP_2 is parallel to the axis of the parabola; Q_1Q_2 is tangent to the parabola at P_2. One shows that the parabolic segment $P_1P_2P_3R$ has an area two-thirds of the parallelogram $P_1Q_1Q_2P_3$.

Let $\alpha(P_1, P_3)$ denote the area of the triangle $P_1P_2P_3$, which is half the area of the parallelogram. Let $\beta(P_1, P_3)$ denote the area of the parabolic segment. Thus,

$$\alpha(P_1, P_3) < \beta(P_1, P_3) < 2\alpha(P_1, P_3).$$

Let A be the midpoint of P_1P_2 with CA parallel to the axis of the parabola. The lemma implies that

$$\frac{CB}{Q_1P_1} = \frac{CP_2^2}{Q_1P_2^2} = \frac{1}{4}.$$

Since $CA = \frac{1}{2}Q_1P_1$, this yields $AB = \frac{1}{4}Q_1P_1$, and hence the triangle P_1BP_2 has area one-fourth that of $P_1Q_1P_2$. Hence $\alpha(P_1P_2) = \frac{1}{8}\alpha(P_1P_3)$. We also have

$$\alpha(P_1, P_3) + \alpha(P_1, P_2) + \alpha(P_2, P_3) < \beta(P_1, P_3)$$

$$< \alpha(P_1, P_3) + 2\alpha(P_1, P_2) + 2\alpha(P_2, P_3)$$

and hence

$$\alpha(P_1, P_3)(1 + \tfrac{1}{4}) < \alpha(P_1, P_3) < \alpha(P_1, P_3)(1 + \tfrac{1}{2}).$$

A process of continued subdivision yields

$$\alpha(P_1, P_3)\left(1+\frac{1}{4}+\cdots+\frac{1}{4^n}\right)<\beta(P_1, P_3)<\alpha(P_1, P_3)\left(1+\frac{1}{4}+\cdots+\frac{1}{4^{n-1}}+\frac{2}{4^n}\right)$$

or

$$\alpha(P_1, P_3)\left(\frac{4}{3}-\frac{1}{3\cdot 4^n}\right)<\beta(P_1, P_3)<\alpha(P_1, P_3)\left(\frac{4}{3}+\frac{2}{3\cdot 4^n}\right).$$

Hence $\beta(P_1, P_3)=\frac{4}{3}\alpha(P_1, P_3)$.

For further discussion of the material in this section, see Dijksterhuis,[4] Morrow,[7] Neugebauer,[8] Neugebauer and Satz,[9] Peet,[10] Struve,[11] and Van der Waerden.[12]

Exercises

4.1. Consider the problem of expressing $2/(2k+1)$ as a sum of fractions with numerator one and different denominators and choosing the expansion in which the maximum denominator is least.

4.2. If a quadrilateral has four sides, a, b, c, d, an ancient formula for the area is $(a+c)(b+d)/4$, where a and c are opposite sides. Show that this formula yields an overestimate in all cases except where the quadrilateral is a rectangle.

4.3. A regular polygon is one that is equilateral and equiangular. The constructions of the regular triangle, quadrilateral, and hexagon are quite straightforward. The construction of the regular pentagon, or five-sided figure, is elementary but more involved.

(a) Suppose given a length a, one constructs a triangle with angles $\alpha=\pi/5$, 2α, 2α and base a. One can then construct the pentagon by a simple compass procedure. (See Figure 4.7.)

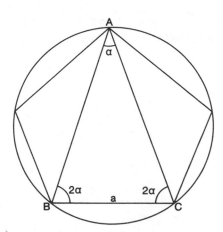

Figure 4.7. The pentagon.

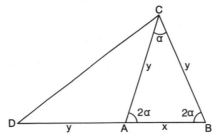

Figure 4.8. Triangle construction for the pentagon.

(b) The construction of such a triangle is clearly the problem of determining the ratio of the side y to the base x. Consider Figure 4.8. Let ABC be a triangle with angles α, 2α, 2α and base $x = AB$. Extend AB to D so that $AD = y$ and complete the triangle DCB. One can show that DCB is similar to ABC and thus

$$\frac{y+x}{y} = \frac{y}{x}.$$

This implies $y/x = \frac{1}{2} + \frac{1}{2}\sqrt{5}$. Given x, one can construct a y in this ratio, by means of a suitable rectangle.

4.4. The regular septagon (seven-sided regular polygon) is not constructible by ruler-and-compass construction.

(a) Let $ABCDEFG$ be a regular septagon (Figure 4.9). Draw the chords AE, AD and the chord GC intersecting AE and AD at P and Q, respectively. One can show that the triangle APQ is a triangle with side AQ equal to one side of the regular septagon and having angles $\alpha = \pi/7$, 2α, and 4α. One can proceed along the following lines:

    (1) Draw AG and AC.
    (2) One shows $\angle EAD = \alpha$, $\angle DAC = \alpha$, $\angle GCA = \alpha$.
    (3) Also $\angle AQP = \angle QAC + \angle QCA = 2\alpha$.
    (4) Also $\angle GAE = 2\alpha = \angle AGC$, $\angle APQ = 4\alpha$.
    (5) $AG = AQ$.

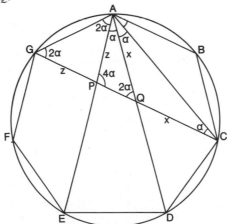

Figure 4.9. The septagon.

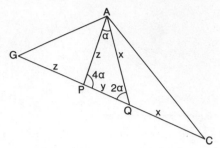

Figure 4.10. Septagon relations.

(b) Suppose that given a side x we can construct a triangle with angles α, 2α, and 4α opposite x (Figure 4.10). Then we can construct a regular septagon of side x. Let the sides opposite α and 2α have length y and z, respectively. One proceeds:

(1) $\alpha = \pi/7$. We extend PQ to G and C, with $GP = z$ and $QC = x$.
(2) Draw AG and AC. $\angle QAC = \angle QCA = \frac{1}{2}\angle PQA = \alpha$.
(3) $\angle GAP = \angle AGP = \frac{1}{2}\angle APQ = 2\alpha$.
(4) $\angle GAC = 4\alpha$, $GA = AQ$.
(5) Pass the circle through G, A, and C. A regular septagon can be constructed by bisecting $\angle AGC$ and doubly bisecting $\angle GAC$. Its side will be $GA = AQ$.

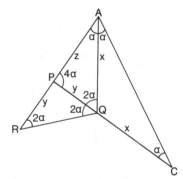

Figure 4.11. Constructive relations.

(c) Given APQ (Figure 4.11), proceed as in (b).
(1) Extend PQ to C, with $QC = x$.
(2) $\angle QAC = \angle QCA = \frac{1}{2}\angle AQP = \alpha$.
(3) $\angle APC$ is similar to $\angle AQP$ and

$$\frac{x+y}{z} = \frac{z}{y}.$$

(4) Extend AP to R with $PR = PQ = y$.
(5) The triangle $\angle RAQ$ is similar to $\angle PAQ$ and thus

$$\frac{y+z}{x} = \frac{x}{z}.$$

(6) We have $xy + y^2 = z^2$ and $yz + z^2 = x^2$. Eliminating y yields (for $x = 1$)

$$z^3 + 2z^2 - z - 1 = 0.$$

(7) The equation (6) on z is irreducible in the rationals. Hence, its group is of order 3 or 6, and hence the root cannot be constructed by ruler and compass.

4.5. Consider the generalization of the preceding discussion to regular polygons of sides 11 and 13.

4.6. The construction of the regular solids is given by Euclid in Book XIII. An analogous problem that provides considerable insight into the geometric relations involved is to determine a plane layout that by suitable cutting and creasing can be assembled into the surface of the solid. Duplicate faces are provided for pasting the figure together. Figure 4.12 illustrates this for the tetrahedron. Figure 4.13 indicates an approach to this problem for the dodecahedron.

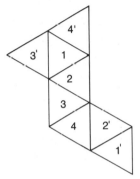

Figure 4.12. Plane layout for tetrahedron with overlap.

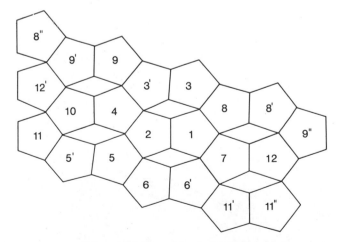

Figure 4.13. Layout for dodecahedron.

4.7. Show that if x is the length of an edge of the dodechahedron and $y = \frac{1}{2}(1 + \sqrt{5})x$, then one can inscribe a cube of edge y in the dodecahedron so that the vertices of the cube coincide with vertices of the dodecahedron. This permits one to determine the radius of the circumscribed sphere.

4.8. Show that a dodecahedron can be inscribed in an icosahedron so that the midpoints of the faces of the icosahedron are vertices of the dodecahedron. What is the ratio of the edges of the figures?

4.9. Establish the symptom for the circle.

4.10. Obtain the symptom for an ellipse. Consider Figure 4.14. The focus of the ellipse is the intersection, C, of the axis, AO, of the cone with the plane, c, of the conic section. The axis of the ellipse is the chord, through C, perpendicular to the intersection,

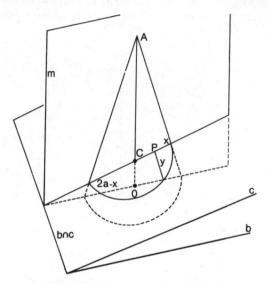

Figure 4.14. Symptom of a conic section.

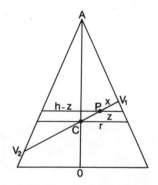

Figure 4.15. Cross section in the plane m.

$b \cap c$. Let P be a point that divides the axis into segments x and $2a - x$, and let y be the perpendicular half chord. Let m be the plane that contains AO and CP. Let Figure 4.15 be in this plane. If we take the plane parallel to the base through P, its intersection with the cone is a circle and we obtain $y^2 = z(h - z)$. One can also show

$$\frac{z(h - z)}{x(2a - x)} = \frac{r^2}{V_1 C \cdot V_2 C} = \alpha < 1,$$

and this yields the symptom.

4.11. Obtain the symptom of the parabola. Consider Figure 4.16. If the plane c of the conic section is parallel to the element AQ, we take the plane m to contain AQ and AC. As before, $y^2 = zr = r^2 x / VC$.

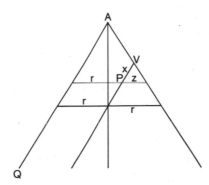

Figure 4.16. Cross section for the parabola.

4.12. Obtain the symptom for the hyperbola. Consider Figure 4.17. This figure is in the plane m, containing AO and perpendicular to the plane c of the conic section. As before, one obtains

$$y^2 - z(h - z) = \frac{r^2}{V_1 C \cdot V_2 C} \, x(x + V_1 V_2).$$

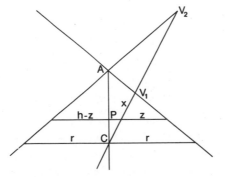

Figure 4.17. Cross section for the hyperbola.

4.13. The procedure for obtaining conjugate axes for a hyperbola involves both the locus of the symptom $y^2 - \alpha x^2 = \alpha a^2$ and the conjugate hyperbola $y^2 - \alpha x^2 = -\alpha a^2$. Thus, $(x_1 \ y_1)$ is chosen on one locus and (x_2, y_2) on the conjugate locus with $y_1 y_2 - \alpha x_1 x_2 = 0$. One can obtain the condition $s^2 = t^2 + 1$ and a corresponding procedure for obtaining tangents.

4.14. Show that the moment of the triangle $P_1 P_2 Q$ around $P_2 Q$ equals the volume of a tetrahedron with base $P_1 P_2 Q$ and altitude equal to $P_2 E$. See Figure 4.4.

4.15. Show that $\frac{1}{3}(\frac{1}{2}QP_1 + CA)/GE^2$ is the volume of the wedge shown in Figure 4.4. This expression can be used to obtain the formulas (m_1) and (m_2).

4.16. The principle of Cavalieri states that if solids have bases in the same plane and if the areas of intersections with planes parallel to the base are equal, then the volumes are equal. Thus, in Figure 4.18 one can show that the intersections of a plane parallel to the base of a hemisphere and cone have a total area equal to the intersection with a cylinder, i.e.,

$$\pi y^2 + \pi x^2 = \pi r^2,$$

and this will yield the volume of the hemisphere, assuming one knows the volume of the cone and cylinder. This is one step in a sequence of comparisons that permits one to obtain volumes on an elementary basis, i.e., without using antiderivatives. Conditions for the equality of volumes in the case of prisms and also in the case of pyramids are established, and these conditions permit comparison with the rectangular parallelepiped by linear and planar constructions and results in the usual formulas for the prism and pyramid. The volume of a cone is obtainable by an exhaustion process. The principle of Cavalieri can be considered to be a volume axiom for this process.

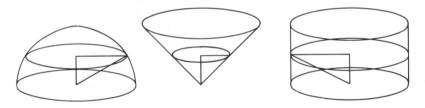

Figure 4.18. Volume comparison.

4.17. Compare the area of a spherical zone with the corresponding band on the circumscribed cylinder with same axis as the zone.

4.18. Since the range of problems that can be solved by ruler-and-compass constructions was limited, the ancient geometers invented mechanical devices to perform constructions equivalent to solving cubic and higher-degree algebraic equations. One such problem was to find the double mean proportional between two line segments, say, a and d. The most straightforward geometrical representation of such a situation is given by the diagram of a triangle APT (Figure 4.19), with lines BQ, CR, and DS parallel to the base AP and the lines AQ, BR, and CS parallel to each other. For such a construction

$$\frac{a}{b} = \frac{AT}{BT} = \frac{QT}{RT} = \frac{b}{c} = \frac{BT}{CT} = \frac{RT}{ST} = \frac{c}{d}.$$

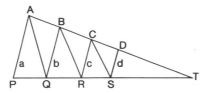

Figure 4.19. Double mean proportional.

Thus, if we let $\rho = a/b = b/c = c/d$, $\rho^3 = a/d$. Mechanical devices involving rods and sliding collars are readily devised that realize the geometrical restraints in our figure. But if these are to be used in a given context, there must be a careful mechanical analysis that must be based on a mathematical analysis. Suppose the lengths a and d, the point P, and the line PT are specified. The triangle APT can then be determined by two further parameters. With $\angle APT$ and the length d given, the triangle DST is determined and so is the rest of the figure. Thus, the figure has two degrees of freedom. Describe mechanisms with varying degrees of freedom and how practical limitations reduce the degree of freedom. For example, input or output may have to be along certain lines.

4.19. One famous ancient problem was trisecting an angle. A diagram that permits a trisection is shown in Figure 4.20. A circle is drawn and the angle to be trisected is realized as the central angle, $\theta = \angle AOB$. One extends AO beyond the circle. One then constructs the desired angle $\angle BRA$ by turn·ᵍ a line BR around B until the segment RS equals the radius r of the circle. If AOP is a diameter and we start from the initial position BP and move the intersection point R out, the segment RS will increase from zero. One can readily show by means of isosceles triangles and the theorem that an exterior angle is the sum of the alternate interior angles that $\theta = 3\alpha$. Discuss mechanisms for bisecting and trisecting angles.

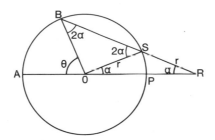

Figure 4.20. Angle trisection.

4.20. List the propositions in Euclid that are incorrect.

4.21. Obtain a plane layout for the icosahedron (see Exercise 4.6).

4.22. What is the value of the dihedral angle between the faces of the various regular solids? For the inscribed sphere suppose a face is tangent at a pole. For the various solids what would be the latitude of the vertices for this face? What can one say about the longitude?

4.23. An Archimedean solid is a convex polyhedron having regular polygons as faces, equal edges, and the configurations around the vertices all congruent. Determine the set of these. What about plane layouts for each?

References

1. Bochner, Salomon, *The Role of Mathematics in the Rise of Science*, Princeton University Press, Princeton, New Jersey (1966).
2. Chace, A. B. *et al.*, *The Rhind Mathematical Papyrus* (2 vols.), Mathematical Association of America, Oberlin, Ohio (1929).
3. De Wulf, Maurice, *The System of Thomas Aquinas* (reprint), Dover Publications Inc., New York (1959).
4. Dijksterhuis, E.G., *Archimedes* (English translation), Ejnar Munksgaard, Copenhagen (1956).
5. Heath, Sir Thomas L., *The Thirteen Books of Euclid's Elements* (reprint), 2nd edition, Dover Publications Inc., New York (1956).
6. Jowett, B., *Plato, "The Republic and Other Works,"* Doubleday and Company, Garden City, N.Y.
7. Morrow, G. R., *Proclus: A Commentary on the First Book of Euclid's Elements*, Princeton University Press, Princeton, New Jersey (1970).
8. Neugebauer, O., *The Exact Sciences in Antiquity* (reprint), Dover Publications, Inc., New York (1969).
9. Neugebauer, O., and Satz, A., *Mathematical Cuneiform Texts*, American Oriental Series, Vol. 29 (1945); published jointly by the American Oriental Society and the American Schools of Oriental Research, New Haven, Connecticut.
10. Peet, T. E., Mathematics in ancient Egypt, *Bull. John Rylands Library*, **15**(2) (July 1931), Manchester University Press, Manchester, England.
11. Struve, W. W., Mathematische Papyrus der staatlichen Museum der Schoenen Künste, *Quellen und Studien zur Geschichte der Mathematik*, Vol. 1, Part A, J. Springer, Berlin (1930).
12. Van der Waerden, B. L., *Science Awakening*, P. Noordhoff, Ltd., Grönigen, Holland (1954).

5

Transition
and Developments

5.1. Algebra

It will be recalled that Sumerian mathematics involved problems for which a solution was given in algorithmic form. The problem was stated in the form of a specific situation with definite numbers, and the procedure for obtaining the solution was a sequence of arithmetic operations. The arithmetic character is in contrast with the "geometric algebra" of Euclid.

Diophantus was a mathematician of Alexandria who wrote a treatise consisting of a sequence of essentially arithmetical problems (see Ver Eeke[30] or Heath[10]). It is suspected that this work is a development of Babylonian mathematics, but certainly it is more sophisticated. Let us consider one of the problems, i.e., 27 in Book I of the *Arithmetica* of Diophantus (the following corresponds, in general, to the modern language used in the French translation of Ver Eeke; the translation in Heath follows the more cryptic original):

> To find two numbers such that their sum and product correspond to given numbers. It is necessary that the square of half the sum less the product be a square....
>
> Suppose then that the sum of the numbers be 20 units and the product be 96 units.
>
> Let the difference be 2σ. Then since the sum is twenty units if we divide this into two equal parts, each of these parts will be half of the sum or 10 units. Then if we add to one of the parts and subtract from the other part, one half of the difference, σ, one will obtain again that the sum of the numbers is 20 units and the difference is 2σ. Consequently the largest of the two numbers is 10 plus σ and the smallest of the numbers is 10 less than σ.
>
> It is required that the product of these numbers constitute 96 units. Their product is 100 less σ^2 which we equate to 96 and $\sigma = 2$. Consequently, the larger number is 12 and the smaller is 8 and these numbers satisfy the requirements.

This problem is elementary and still well within the capabilities of Babylonian mathematics. Nevertheless it has distinguishing characteristics that represent considerable intellectual development:

(a) The numbers are dissociated from any enumeration or measurements of a specific situation such as occurred in the older mathematics or from a geometric interpretation as· in Euclid. Instead the numbers are expressed as a quantity of an abstract unit, $\overset{\circ}{M}$, in the classical Greek system. This system is a nonplace decimal system in which different letters are used for units, tens, and hundreds. Thus 2, 20, and 22 would be represented $\overset{\circ}{M}\beta$, $\overset{\circ}{M}\kappa$, and $\overset{\circ}{M}\kappa\beta$, respectively.

(b) The discussion is in terms of properties of numbers. It is a running prose argument that one can readily translate into equation form. Thus, it is equivalent to modern elementary algebra.

(c) The numbers are rational and positive so that one must require that a number must be a square of a rational if one is to take a square root. Negative and imaginary solutions to problems are ignored. Many such problems of Diophantus are equivalent to finding integers or classes of integers that satisfy polynomial relations, and the modern term "Diophantine equations" refers to such situations

(d) One development was the use of a symbol, σ, for an unknown quantity. Presumably σ stands for $\alpha\rho\iota\theta\mu o\sigma$, the word for number. The various powers of σ were also indicated by symbols. The square of σ is represented by Δ^v for $\delta v v \alpha\mu\iota\sigma$ (power), the cube of σ by K^v for "$\kappa v \beta o\sigma$" (cube), and the fourth, fifth, and sixth powers, by $\Delta^v\Delta^v$, $\Delta^v K^v$, and $K^v K^v$, respectively. There was also a way of representing the reciprocals of these powers.

(e) The problem is stated using the term "given numbers." But in the procedure this term is replaced by specific values, and the development is the same relative to these as in the previous algorithmic mathematics.

The work of Diophantus continued to have considerable influence on subsequent mathematics and mathematicians. For example, Fermat's interest in mathematics was awakened by the edition compiled by Bachet, and Fermat's famous "last theorem" was stated in the margin of this work. From the collapse of the Roman Empire in the West until the fall of Constantinople in A.D.1452, mathematics was mostly in the hands of Arabian and Eastern scholars who emphasized the "problem tradition" of mathematics. One important development was that of decimal arithmetic using a place notation. The standard histories of mathematics offer many fascinating details.

During the Renaissance there was a tremendous European interest in mathematics, and this interest produced a more sophisticated and effective algebra. The symbols $+$, $-$, $=$, $\times$, $\div$, and $\sqrt{}$ were introduced in the sixteenth and early seventeenth centuries. In the work of Vieta (1591) there appears a logical development of algebra based on the appropriate theorems of Euclid. Vieta regarded algebra in which operations are on letters as representing a higher form of mathematics without hypotheses such as Proclus desired. Descartes introduced (1637) the notation x, y, and z for the unknown quantities.

The cubic and quartic equations were solved by Italian mathematicians of the fifteenth century (see Smith[25]). A cubic equation would be expressed at that time as "cub^{s}p; 6 reb^s aeglis 20." The superscript s refers to the unknown and the expression is shorthand for "the cube plus 6 times the thing itself equals 20." The method used is applicable to a general cubic equation in the form $x^3 + px = q$, although it was stated for specific values of p and q. Consider the identity in u and v

$$(u-v)^3 + 3uv(u-v) = u^3 - v^3.$$

If we find a u and v such that $3uv = p$, $q = u^3 - v^3$, then $x = u - v$ is a solution. But if we let $a = u^3$, $b = v^3$, we have $a - b = q$ and $ab = p^3/27$, and we can readily find a and b.

The solution of the quartic is dependent on solving an intermediate cubic. Suppose we wish to solve $x^4 + ax^2 + bx + c = 0$. We can write this in the form

$$(x^2 + r)^2 = sx + t$$

for $r = a/2$, $s = -b$, $t = -c + a^2/4$. Let us now add a quantity y to $x^2 + r$. Then

$$(x^2 + r + y)^2 = 2yx^2 + sx + t + 2ry + y^2.$$

The right-hand side is in the form $2y(x + A)^2$ for $A = s/4y$ provided

$$y^3 + 2ry^2 + ty - s^2/8 = 0.$$

If we solve this cubic equation, we can express our given equation in the form

$$(x^2 + r + y)^2 = 2y(x + A)^2.$$

We can now extract the square root of each side and obtain a quadratic in x.

Clearly these procedures are based simply on the binomial theorem for the second and third powers. The formulas for arithmetic and geometric series were of course known. There are also formulas for sums of powers of integers such as

$$1 + 2^2 + \cdots + n^2 = (n+1)(2n+1)n/6,$$

which are readily proven by induction on n.

This European development of algebra presented arguments using an increasingly flexible notation in which the manipulation of equations appears to have a logical significance equivalent to the arguments of geometry. Nevertheless, the classical approach in which numbers or the quantitative aspects of magnitudes have an adjectival character rather than an objective character was still dominant, and arguments could be considered rigorous only if they could be referred to Euclidean or Archimedean axioms.

But the increased facility in algebra was itself significant. The use of the symptoms for the conic equations led naturally to analytic geometry and new ways for finding tangents to curves. The formulas in which n was permitted to be indefinitely large led naturally to the idea of an infinite sequence. These formulas also provided methods of finding areas that were a form or at least a precursor of the integral calculus.

For further discussion of the material in this section, see Smith,[25] Struik,[26] and Ver Eeke.[30]

5.2. Non-Euclidean Geometry

For almost twenty centuries the fifth postulate of Euclid was a challenge to those who believed that it was unnecessary and a consequence of the remaining assumptions of geometry. But this effort was climaxed by the recognition that the postulates of Euclid correspond to a precise analysis of the relationship between points and lines in what is now referred to as the Euclidean plane and that variations on the fifth postulate yield other, "non-Euclidean," geometries. It is a remarkable justification of Euclid.

The notion of an angle at a given vertex occurs at the very beginning of Euclid, as does the configuration of linear angles at a given vertex. Postulate 4, which is critical, states that all right angles are equal. It implies immediately the equality of vertical angles, but this would also follow if one assumed that all the right angles at any one point are equal. The significance of Postulate 4 can be illuminated by considering cases, in which right angles at different points are different. For example, the surface of one nappe of a cone has a geometry in which the configuration of lines through a given point is quite similar to that for a point in the Euclidean plane except for the vertex. A right angle at the vertex is equal to an acute angle at any other point. (A right angle is half a straight angle. To define straight lines for points other than the vertex, we would use the fact that the surface of a cone can be laid out so that any sufficiently small piece not containing the vertex is flat. A straight line through the vertex would be a configuration of two coplanar elements.) On the other hand, many Riemann surfaces have exceptional points at which a right angle is equal to a straight angle at other points. The concept of a tangent plane permits one to compare the angle and line configurations of points on quite general surfaces.

Postulate 5 permits one to compare point configurations at different points. The line joining two points can be used to orient the angular configuration associated with each point. The interplay of incidence relations for the rays originating at the points and the angles of these rays with the reference line is a fundamental aspect of the geometry.

For example, one can distinguish between three kinds of geometry by means of the Saccheri quadrilateral. Let us start with a right angle ABC, and at C erect another right angle BCD, and at D erect the right angle CDA. If the right angles are "interior" right angles, Postulate 5 insures that DAB is a right angle. But one can also consistently require it to be an acute angle, which yields hyperbolic geometry, or one can require it to be obtuse, as in spherical and elliptical geometry.

In differential geometry, the angular configuration at a point consists of the tangent lines of curves through the point. The usual bilinear form yields the cosines of angles. An important idea here is the way in which rays at one point are associated with rays at another. This is referred to as "parallel displacement." Thus, even in this new sophisticated geometry, the basic analysis of Euclid is still valid.

The process by which non-Euclidean geometries were discovered is, of course, well known. Its major significance is in the fact that it established that geometry was not an ideal abstraction of spatial experience corresponding to a higher form of truth. For if this were so there would not be this ambiguity, which can only be resolved by experiment. Mathematics provides invented geometrical constructions, some of which may be associated with experience in the form of a mathematical theory for spatial experience.

For further discussion of the material in this section, see Carslaw[2] and Manning.[15]

5.3. Geometric Developments

The result of introducing Cartesian coordinates into geometry is familiar to students of mathematics. Having chosen a unit length it is natural to associate the possible geometric ratios with the points on the line. This yields the equivalent of real numbers, including negative numbers, and this was an important addition to algebra. The geometric interpretation of complex numbers was also important in obtaining acceptance (see Smith[24]). The equivalence of pairs and triplets of real numbers with two- and three-dimensional Euclidean space meant that geometric arguments and constructions have equivalent algebraic equations and manipulations. When this analytic structure is generalized to n dimensions, it is of course a form of analysis expressed in a geometric terminology.

Classical geometry contained a considerable body of knowledge that was available for this transition. The use of algebra greatly expanded the possibilities for various geometric notions. For example, many more surfaces could be considered. Analytic geometry was also intimately related to the development of the calculus, and the notion of limits permitted general procedures for taking tangents and tangent planes. But perhaps the most basic aspect of the use of numerical models for geometry was their incorporation into the "main line" of mathematics, in which arguments are based on set-theoretical logic and which, of course, includes the real numbers.

The self-imposed requirement of classical mathematics that the logical discussion should be based on explicitly stated "first principles" and that only they be used was never completely fulfilled. As long as the emphasis was on reasoning involving diagrams these discrepancies were not often considered, and indeed for a long time there was not the rigorous mathematical discipline that would be sensitive to such logical inadequacies. But after analysis had been established on a set-theoretical logical basis, there was considerable interest in the corresponding formulation of the classical example of axiomatic development by distinguished mathematicians such as Schur and Hilbert.

Actually, there was a considerable logical refinement in regard to various geometrical notions such as incidence relations and the order of points on a line. Consequently, it was possible to formulate not only Euclidean geometry on a precise basis but also a variety of other geometries by changing or omitting axioms. The general resurgence of interest in mathematics, which began in the sixteenth century, had a rather small but concomitant development in classical diagrammatic geometry, which reached a striking climax in the nineteenth century.

Thus, the present total logical development called "synthetic geometry" greatly exceeds the classical development and has a logical structure satisfying the same rigorous standards as any other part of modern mathematics. There is a tendency to consider this synthetic geometry as of no practical significance, but it is not unusual to find interesting and fascinating applications. But the type of logical reasoning that structures modern synthetic geometry is highly significant for applications. The normal conceptual procedure is in diagrammatic forms and thus has the appeal and imaginative capabilities of traditional geometry. But these constructions and deductions must be disciplined so that each step can be precisely analyzed into the basic incidence and order axioms.

In the set-theoretical logical form, a geometry is considered to be a conglomerate of sets of various kinds of objects with relations between objects in different sets or even in the same set. Thus, for an incidence geometry in a plane one has two kinds of objects—points and lines—and the relations of "point on a line" and "order of points on a line." There would also be a number of constructive axioms stating the existence of examples of such relationships under various hypotheses.

For such conceptual conglomerates there are many metaprocedures, such as monomorphisms, which are one-to-one correspondences of one conglomerate into another that preserve the internal structure, or isomorphisms, which are monomorphisms onto. A geometry can be shown to be free from contradictions if there is an "analytic model," i.e., an isomorphic image consisting of numerical concepts. This assumes that the number concepts themselves do not imply a contradiction. One may also consider monomorphic images in Euclidean geometry that one may be willing to accept as consistent. Models can also be used to show the independence of a specific axiom from others in a set. The concept of numerical limit in relation to analytic models permits one to establish in a consistent way the notion of tangency and a number of other limiting relations between analytic models.

For further discussion of the material in this section, see Daus,[4] Forder,[8],[9] Hilbert,[12] and Lines.[14]

5.4. Geometry and Group Theory

It will be recalled that Proclus described the "generality" inherent in geometrical reasoning as due to the assumption that reasoning with a typical example would be valid for all examples of the same type. In the argument only properties explicitly connected with the type are to be used. But if we are looking at geometry as a whole, this raises the question as to what properties are available to form "types." For rectilinear figures the basic properties are equality of angles or equality of line segments. When figures have certain combinations of both these properties we have congruence, and when only angular properties appear we have similarity.

This suggests that one looks more closely at the notions of equality of angles and line segments and congruence. Two figures are said to be congruent if there is equality of corresponding angles and line segments. But equality of angles or line segments would usually be interpreted to mean that if one "moved" one figure to coincide in part with another figure, the part in question would completely coincide. But such a "motion" can be considered as a construction of the moved figure in a new position. This construction produces for any given point in the plane a corresponding "moved" point and thus is a "point transformation." The reader will immediately recognize that the set of these point transformations constitute a group for the congruence case, and this is also true for similarity situations. The equality properties are invariants of the figures under these groups of point transformations. Thus, a geometry is associated with a group of point transformations, and the theorems become statements about invariance.

In the Euclidean plane the group of motions associated with congruence contains the parallel translations, the rotations around points, and the inversions in lines. For similarity one must add the homothetic transformations, which are the expansions or contractions in a fixed ratio from a given point. There is, however, an even larger group of transformations, which preserves the angular configuration at each point in the sense of preserving the angles between tangent lines. This group of conformal transformations includes the inversions in circles and is explored in terms of mappings of the complex plane.

We can reverse this process and start with a group of point transformations and concern ourselves with the invariants and the corresponding inferences we could make relative to figures. The group of point transformations can be expressed either analytically or by synthetic constructions. This represents a modern metamorphism of geometry that has become highly significant in quantum mechanics and as the theory of Lie groups in mathematics.

The notion of a group of transformations under which certain properties are invariant can be associated with the mathematical theory of a science. If one is willing to assume on the basis of experimental evidence that such a group exists, one has a much more powerful mathematical machinery for the purpose of inference. Actually, in physics, for instance, one seems to have a complex of groups that of course corresponds to a complex of geometries.

How does one infer the existence and specifications of such a group or groups on the basis of experimental evidence? One might expect that this is some general inductive process of inferring the laws of a theory from experimental evidence. However, what actually happened in one instance in physics is instructive. Originally Newtonian physics was set up on the basis of the "Galilean group" in space–time. A transformation in the Galilean group is determined by a change of a system of coordinates in space–time. The relevant system of coordinates consists of a Cartesian system in space and an additional time coordinate axis. Changes includes the Euclidean changes in space for such systems with the time coordinate unchanged, as well as a subgroup of "uniform" motion changes given by

$$x' = x - vt, \qquad y' = y, \qquad z' = z, \qquad t' = t,$$

where v is a group parameter. The full Galilean group is generated by the transformations mentioned.

For a considerable range of mechanical phenomena this formulation of Newtonian physics on the basis of the Galilean group was satisfactory.

However, Maxwell's equations for electromagnetic phenomena are not invariant under the Galilean group, but under the Lorentz group in which the above-mentioned one-parameter group is replaced by

$$x' = [1 - (v/c)^2]^{-1/2}(x - vt), \qquad y' = y, \qquad z' = z,$$
$$t' = [1 - (v/c)^2]^{-1/2}(t - vx/c^2)$$

and the spatial transformations with t unchanged remain the same.

Thus, the Newtonian laws of mechanics and the phenomena of electromagnetism are associated with different space–time geometries, and this cannot be right, since it means we can appear to have changes in one set of phenomena but not in the other by a mere change of coordinate systems. Electromagnetism has had an extensive experimental verification of its Lorentz invariance. The most effective test of Newtonian mechanics is in the detailed motion of the solar system. In general the velocities in the solar system are such that in the range that is experimentally available the difference between corresponding transformations in the two groups is practically indistinguishable. However, there is a small but measurable discrepancy between the Newtonian prediction for the motion of the planet Mercury and the actual motion. But if one uses a mechanics invariant under the Lorentz group, this discrepancy disappears. Thus, the Lorentz group geometry should be used for both sets of phenomena.

Thus the choice of groups was a matter of adjusting to experience not to a large-scale inductive process. The search for an appropriate group is clearly equivalent to searching for the correct axioms of a geometry. A geometry may have a simpler geometry as a limit when certain parameters approach zero. The expansion of an area of experience may require one to go from a simpler to a more complex geometry, represented by a different choice of a group of transformations or modification of a set of axioms. This is not a matter of arranging a large number of facts into a pattern discovered inductively, but rather adjusting a pattern to conform to a wider range of experience.

For further discussion of the material in this section, see Bierberbach,[1] Daus,[4] and Einstein et al.[7]

5.5. Arithmetic

Computational procedures were referred to in classical times as "logistics" as distinct from the more intellectual consideration of the properties of numbers, which was referred to as "arithmetic." Arithmetic in the form of computation is associated with two distinct areas of application. One of these is scientific or technical, and in the ancient world this was mainly astronomy. Astronomical tables and calculations were performed in a hexagesimal system in which the numbers in specific places were expressed in the Greek decimal system, that is, with letters for different multiples of ten and for different digits.

But arithmetic is also part of trade, tax collecting, and military logistics, and it is more than likely that these applications influenced its development. The use of hexagesimal fractions was probably originally associated with coinage, and it is also possible that the place system originated with counting boards.

Counting boards were intended to perform the arithmetic associated with business or government transactions. Numbers were represented by means of markers placed in rows ruled on the board. Each row corresponded to a place in a place system. The number of markers in a row could correspond to the digit in this place, or the markers could have symbols on them to indicate their value, or colored markers of different values could be used. The arithmetic operations were performed in a manner similar to those on the abacus (see Moon[17]).

Counting boards were widespread until paper became inexpensive enough to permit its use for computation. The word "calculus" means a little stone, i.e., a marker. A variation of the counting board was a sandbox with divisions in which marks could be made in the sand with a finger. The abacus in its modern form was a further development.

The development of the decimal place system for representing integers, with a digit zero, was due to Hindu mathematicians and was introduced to the West through Arabian channels. Decimal integers could be used with fractions such as twelfths in a wide range of applications, but decimalization was climaxed by the use of decimal fractions in the work of Stevin (1585), and this permitted the development of our usual arithmetic in the next 150 years.

A calculating machine was designed by Pascal, but essential improvements were due to Leibnitz, whose logical design was used in calculators for many years. Rotatory calculators were of great practical significance until the 1950s and used a completely decimal notation. But advances in electronic circuitry led first to large-scale data processing, which dominated business and government, and then to small devices of remarkable capability. Electronic circuitry is better adapted to use the binary or radix two system rather than the decimal system. Boolean algebra is readily represented, as are two-state circuits.

For further discussion of the material in this section, see Moon,[17] Murray,[19] and Struik.[26]

5.6. The Celestial Sphere

Astronomy and mathematics interacted on each other in many significant ways, and our modern exact sciences are lineal descendants of this interaction. For most applied mathematicians even the more elementary technical relations between these subjects are important, and these relations are readily available in such books as that by Smart[23] Students interested in applied mathematics will certainly find in this area valuable and fascinating ways to enhance their facility with three-dimensional geometry. However, our immediate concern will be with certain aspects involved in the development of mathematical understanding.

The distances of the fixed stars from the sun are enormous compared with the width of the earth's orbit. The angular displacement of even the nearest fixed stars is about 1.5″ of arc and can be observed only by telescopic photography. Thus, the stars have fixed angular relations when observed from earth, and these angular distances seem to be constant in time and space. Thus, one may consider the stars to be on a "celestial sphere" of indefinitely large radius.

Due to the rotation of the earth, this whole sphere appears to revolve in a little less than a day. After the twilight that follows sunset, a certain part of this sphere is visible. This visible part turns westward, with a new portion rising in the east. This motion is a rotation around a fixed line parallel to the earth's axis. The points where this axis intersects the celestial sphere are the poles. In the northern hemisphere we see only one—the North Pole. The North Pole is fixed relative to us as well as on the celestial sphere.

In addition to this fixed axial direction, a plumb line will determine a direction of "straight up" or "straight down." We can also think of a sphere, fixed relative to us, and coincident with the celestial sphere. These two fixed directions determine a plane called the plane of the meridian. This plane intersects the fixed sphere in a great circle called the meridian. The celestial sphere rotates past this great circle westward.

The North Pole corresponds both to a point on the celestial sphere and a point on the fixed sphere. Thus, the great circles for which it is the pole coincide and these are called the equator. Of course, the equator rotates with the celestial sphere when it is considered on the celestial sphere so that its intersection with the meridian moves eastward on the celestial sphere. Another great circle on the fixed sphere is the horizon, which has the vertical direction as its pole. The horizon is mathematically defined; it is not the apparent horizon of observation.

Gravity permits one to readily determine the vertical direction and the horizontal plane. The direction of the North Pole can also be readily determined. These determine a spherical coordinate system in which a longitude variable, the "azimuth," is measured along the horizontal circle from the northern meridian point. The latitude variable is called the elevation. Azimuth as designated as either "east" or "west." The zenith is the point immediately overhead, and the "zenith distance" of a point is the complement of the elevation. With limited resources, these angles—azimuth, elevation, and zenith distance—are most readily measured.

The light of the sun blanks out the stars and the celestial sphere when the sun rises, but in a single day it moves across the sky from east to west on a path similar to that of the stars on the celestial sphere. Thus, it is natural to assume that the motion of the sun is part of the motion of the celestial sphere, and this will agree with casual observation over a few days. But more careful observation will show that the part of the sphere visible at night changes somewhat from night to night as if the celestial sphere were turning faster around the earth than the sun.

One can assume that the sun partakes of the motion of the celestial sphere but moves backward on it one complete revolution a year. Because of this relative motion, the part of the celestial sphere visible at night at different times in the year changes, so that one can plot the stars that appear on this entire globe, except for a cap around the South Pole, which is always below the horizon. The size of this unseen cap depends on latitude. On this completed globe the sun follows a fixed path, which appears at first to be the same from year to year.

This apparent motion of the sun on the celestial sphere is due to the orbital motion of the earth. The mean distance of the sun is about 23,000 times the radius of the earth, so that to an observer at the North Pole and one at the equator, the line of sight to the sun would be parallel to within 8″ of arc. Thus, the sun appears at any instant to have a position on the celestial sphere that is practically independent of the position of the observer. The vector joining the earth to the sun always moves in the orbital plane, and this vector, of course, corresponds to the apparent direction of the sun from the earth. Thus, the apparent motion of the sun is essentially on a plane fixed in regard to the observer. The intersection of this plane with the celestial sphere is the great circle called the ecliptic, the path of the sun on the celestial sphere.

The ecliptic intersects the equator in two points, one of which is associated with the constellation of Aries. To an earthbound observer the ecliptic appears to revolve with the celestial sphere, so that this point of Aries, which always moves on the fixed great circle the equator, corresponds to the position of a hand on a 24-hour clock. If we take the zero hour to be the instant when this point crosses the visible part of the meridian, the sidereal time of day is indicated by the westward arc on the equator from the meridian to the present position of the point in Aries.

On the other hand, this point of Aries is also a fixed reference point on the equator of the celestial sphere. Thus, on the celestial sphere itself, one can set up a spherical coordinate system with one angular coordinate similar to longitude measured eastward along the equator from this point in Aries. This coordinate is called the right ascension of a star position. The equivalent of latitude is called the declination and is positive toward the North Pole. This system of coordinates is called "equatorial." There is a similar system called "zodiacal" with "celestial longitude" measured eastward along the ecliptic from the point in Aries and "celestial latitude" positive on the North-Pole-side of the ecliptic.

For further discussion of the material in this section, see Smart.[24]

5.7. The Motion of the Sun

For people living in the temperate zone, probably the most obvious celestial phenomena is the seasonal variation in the length of day and the related position of the sun during the day. These depend directly on the declination of the sun. In its motion along the ecliptic, the sun is above the equator during spring and summer and below it during the fall and winter. The dependence of the length of day on the sun's declination, δ, is discussed in Smart[24] (Chapters II and III). The sun has celestial latitude 0, and hence $\sin \delta = \sin \varepsilon \sin \lambda$, where $\varepsilon \simeq 23° 27'$ and λ is the celestial longitude of the sun and should therefore be considered as a function of the time (see Exercise 5.19).

The time between two successive passages of the sun through the point of Aries is practically constant whether measured in terms of the rotation of the earth or by frequency procedures based on atomic phenomena. Thus, λ, the sun's longitude, is an increasing function of the time that differs from a linear function $nt + \varepsilon_0$ in a periodic way, i.e., $\psi(t) = \lambda - nt - \varepsilon_0$ will be zero at $t = 0$, if ε_0 is the corresponding value of λ and at $t = T, 2T, \ldots$, where T is the number of time units in a year. Thus, λ changes by 2π as t changes by T, and $n = 2\pi/T$.

Now consider a polar coordinate system (r, θ) for the orbital plane of the earth centered at the sun and with reference line joining the sun to Aries. One can readily see (Figure 5.1) that $\theta = \lambda + \pi$ (mod 2π). If $\bar{\omega}$ is the angular coordinate of perihelion, the orbital ellipse is given by

$$r = \frac{a(1 - e^2)}{1 + e \cos (\theta - \bar{\omega})},$$

where e is the eccentricity of the ellipse. Then Kepler's second law becomes $r^2 (d\theta/dt) = h = a^2 n (1 - e^2)^{1/2}$ (Smart,[24] p. 100). Let $v = \theta - \bar{\omega}$, $\lambda = v + \bar{\omega} \pm \pi$,

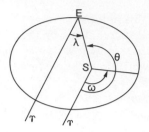

Figure 5.1. The earth's orbit.

where the sign is chosen so that $0 \leqslant \lambda \leqslant 2\pi$. If $\bar{t}$ is time of perihelion, then $v=0$ at $t=\bar{t}$. Then

$$\frac{dv/dt}{(1+e\cos v)^2} = \frac{n}{(1-e^2)^{3/2}}. \tag{1}$$

This equation can be integrated (see Exercise 5.20) into the form

$$v + \phi(v) = n(t-\bar{t}), \tag{2}$$

where

$$\phi(v) = \sum_{n=1}^{\infty} (-1)^n [2(1-e^2)^{1/2} + 1/n] e^{*n} \sin nv \tag{3}$$

for $e^* = e/[(1-e^2)^{1/2}+1]$. Now for the earth's orbit, $e=0.016726$ (1960) and $e^*=0.0083636$, and for the given accuracy we have

$$\phi_E(v) = -0.033452 \sin v + 0.000210 \sin 2v. \tag{4}$$

For the major visible planets in the solar system the eccentricity of the orbits are relatively small numbers. Mercury has an e of about 0.2, and for Mars $e=0.093368$ (Smart,[24] p. 422) and

$$\phi_M(v) = -0.186736 \sin v + 0.006547 \sin 2v - 0.000272 \sin 3v$$
$$+ 0.0000119 \sin 4v - 0.0000005 \sin 5v. \tag{5}$$

(We have retained an extra figure to minimize roundoff.)

Because of the smallness of the coefficients in Equations (4) and (5), it is relatively easy to transform Equation (2) into the form

$$v = n(t-\bar{t}) + \Psi(n(t-\bar{t})). \tag{6}$$

See Exercise 5.21 and 5.22. In particular we have

$$\Psi_E(x) = 0.033451 \sin x + 0.00349 \sin 2x + 0.000003 \sin 3x \tag{7}$$

and

$$\Psi_M(x) = 0.186533 \sin x + 0.010863 \sin 2x + 0.000877 \sin 3x$$
$$+ 0.000081 \sin 4x + 0.000007 \sin 4x. \tag{8}$$

Thus

$$\lambda = n(t - \bar{t}) + \Psi(n(t - \bar{t})) + \bar{\omega} \pm \pi. \tag{9}$$

The table in Smart[24] (p. 422) gives $\bar{\omega}$ and the value of θ, ε_0 at $t = 0$. This permits one to find $n\bar{t}$, since

$$-n\bar{t} + \psi(-n\bar{t}) + \bar{\omega} = \varepsilon_0, \qquad \text{or} \qquad n\bar{t} + \psi(n\bar{t}) = \bar{\omega} - \varepsilon_0. \tag{10}$$

In view of the relation of Equations (2) and (6), this yields

$$n\bar{t} = \bar{\omega} - \varepsilon_0 + \phi(\bar{\omega} - \varepsilon_0) = \tau_0. \tag{11}$$

Thus, if

$$M = nt - \tau_0 + \bar{\omega}_0 = nt + \varepsilon_0 - \phi(\bar{\omega} - \varepsilon_0), \tag{12}$$

$$\lambda = M + \Psi(M - \bar{\omega}_0) \pm \pi. \tag{13}$$

M is a "mean" longitude with a uniform rate of change, and the "true" longitude λ differs from M by $\pm \pi$ and a periodic function Ψ, which has values less than 0.033454 radians, or $2°$. Note that Ψ is always less than the change in M for 2 days.

For further discussion of the material in this section, see Smart.[24]

5.8. Synodic Periods

Except for the moon, the objects in the solar system are at extremely large distances relative to distances available on earth. Venus comes closest to the earth, but if one had a baseline of 1000 kilometers between two observers, their two lines of sight toward Venus would still be parallel to within 5″ of arc at the nearest approach. Both the radius of the earth and the distance to the moon were measured in ancient times by procedures based on simple angle-measuring instruments, and the answers were probably correct to within a few percentage points. But the distances to the sun and the planets were either recognized as unavailable or greatly underestimated.

Thus, direct geometric procedures can only yield angular information for the planets. This information in general deals with periodic phenomena or nearly periodic phenomena. For example, the outer planets are most readily observed when they are precisely opposite the sun on the celestial sphere. The orbital planes of these planets are inclined at about 1 or 2 degrees to the plane of the ecliptic. Thus, the periodic motion of the planet and that of the earth will cause the sun, the earth, and the planet to align in longitude at intervals that can be predicted approximately on the basis of a long-time average. Let M_E and M_P be the M of Equation (12) for the earth and the planet, and suppose that at time $t=0$ both have the value m_0. Then $M_E = (2\pi/T_E)t + m_0$ and $M_P = (2\pi/T_P)t + m_0$. Since M_E changes faster than M_P, the next "mean" alignment will occur when $M_E - M_P = 2\pi$. If this corresponds to a time interval τ, then $1/T_E = 1/T_P + 1/\tau$. If we ignore the difference between the orbital planes, what is really desired is that λ_E and λ_P coincide up to 2π, and the equation becomes

$$M_E + \Psi_E(M_E - \bar{\omega}) = M_P + \Psi_P(M_P - \bar{\omega}) + 2\pi.$$

This will require us to correct the τ value for each individual conjunction. But the relative orbital position of the earth at conjunction determines the relative orbital position of the planet. This implies that the correction is a function of the time of year.

In general, if only nontelescopic instruments are available, immediate angular measurements may not be very accurate, but if careful records are kept, long-time averages may be much more precise, and the phenomenon lends itself to empirical formulation. The modern form of such "empirical formulation" would undoubtedly be a Fourier series. The predictions for the moon are probably the most difficult.

5.9. Babylonian Tables

The integration of this experience into a consistent arithmetical procedure for prediction was an excellent intellectual accomplishment. Presumably this occurred in Babylonia before 300 B.C. (see Neugebauer[20]). The celestial sphere was introduced, and a zone, the "zodiac," around the ecliptic was divided into twelve equal sectors, corresponding to the constellations or "signs of the zodiac"—Aries, Taurus, Gemini, etc. Thus, Aries corresponds to longitude $0°$–$30°$, Taurus to $30°$–$60°$, etc. Babylonian arithmetic was quite adequate to yield prediction tables for years ahead to correspond to observable phenomena. Long-time averages were represented quite precisely, and angular phenomena were represented in a manner quite adequate for observation.

Babylonian astronomy is described by Neugebauer[20] (Chapter V, p. 97). The characteristics described above are well illustrated by his examples (p. 110) of the changes in longitude of the sun in monthly intervals. The first column is a succession of dates corresponding to equal time intervals of a mean synodic month. The second column is the change in the longitude of the sun since the previous date, and the third column is the actual longitude of the sun expressed by degrees within a constellation. Let $(\Delta\lambda)_i$, $i=0, 1, ..., 12$, denote the values in the second column. Thus, this table gives the position of the sun relative to a lunar calendar.

The $(\Delta\lambda)_i$ values vary from month to month but are reasonably close to an average value μ. The $(\Delta\lambda)_i$ values vary between values M and m along a "sawtooth" curve, so that except where the sawtooth changes slope, successive tabulated values of $(\Delta\lambda)_i$ are obtained by adding or subtracting $s=18'$. The period of the sawtooth is one year. The entries, $\Delta\lambda$, are given in degrees, minutes, seconds, and $\frac{1}{60}$ths of a second.

To construct this table, one must know or choose three quantities. One of these is v, the number of synodic months in a year, the result of a long-time averaging process. If μ is expressed in degrees, $\mu=\frac{1}{2}(M+m)=360°/v$. The total variation of the sawtooth function over a year is $2(M-m)$, and if s is the slope of the sawtooth, $sv=2(M-m)$. Thus, v and the choice of the slope s determines M and m. The remaining quantity is a time-phase quantity that determines where the first value of $\Delta\lambda$ in the table is to be taken on the sawtooth. Variations in the choice of slope or phase will introduce relatively small errors in individual values. But an error in the value of v or μ would produce an error that would increase with time.

It is interesting to compare the values used with modern values. The value of μ given is $29.10537°$. The modern value is $29.10675°$ and the difference $0.00138°$ is approximately $5''$ of arc. Similarly the maximum value M is $30.03306°$ and the corresponding modern value is $30.08965°$ with a difference corresponding to $3.4'$. Similarly the m used is $28.17769°$, the calculated minimum is $28.16274°$, with a difference of $0.9'$ of arc. For the modern case, a slope of $18.7'$ would have been chosen. These values indicate maximum errors of a few minutes. If we suppose that the sawtooth should have been replaced by a sine curve, the maximum difference between sawtooth and sine would be $0.39°$ or $23.4'$. This would seem to indicate that while long-range averages may have percentage accuracies of 0.005%, the angular measurements were probably not sensitive to much less than half a degree. The Babylonians also had tables for the motion of the moon and for planetary phenomena.

For further discussion of the material in this section, see Neugebauer.[20]

5.10. Geometric Formulations

Greek mathematicians including Apollonius and Hipparchus set up geometric models for solar system motions using exocentrics and epicycles that gave a better qualitative description. The Ptolemaic system is described in Neugebauer[20] (Appendix 1, p. 191), and there are many references in Van der Waerden.[29] If the orbital motions of the planets were circular and consequently uniform, then the epicyclical description of the angular motions referred to a fixed earth would be precisely correct. Under this assumption the sun would appear to uniformly circle the earth and the inner planets would circle around the sun. For the outer planets this is an apparent motion that is indistinguishable from the actual motion. For example (see Figure 5.2), if one represents the displacement of Mars relative to the sun by a vector $\bar{y}(t)$ and that of the earth relative to the sun by a vector $\bar{x}(t)$, then the angular motion of vector ME is given by $\bar{y} - \bar{x}$. This is identical with a motion obtained as follows. Let C complete the parallelogram with vertices ESM (Figure 5.2). Then $\overleftarrow{CE} = \bar{y}$ and $\overleftarrow{MC} = \bar{x}$ and C revolves around the earth the same way as M does around S, and M revolves around C the way E revolves around S. Thus, ME can be considered as due to the rotation of M given by $\bar{x}$ about moving C given by $\bar{y}$. Thus, if the orbital motions were uniform, an epicyclical description would be valid.

But the orbital motions are not quite uniform even though the eccentricities of the major planets are not large numbers. One can therefore try to "compound" the epicyclical models by describing the nonuniform orbital motions by epicycloids. One can begin with the apparent motion of the sun in longitude, i.e., the equivalent of the earth's orbital motion, which appears in the epicyclical description of every planet.

Let us obtain a first-order fit to a Kepler orbital motion by means of an epicycle. Consider Figure 5.3 and suppose P moves around S in an epicyclical fashion, with CS (the "referrant") moving with radius R and with PC (the eccentric) with radius $r = \rho R$, say, with $\rho < 1$. CS has rotated an amount τ and PC an amount σ relative to SC, and τ and σ are both linear functions of the time. For the earth orbit, we will have $\tau - \sigma$ equal to a constant which we can take as zero. Then θ, the heliocentric longitude, equals $\tau + \alpha$, where $\alpha = \arctan[r \sin \sigma/(R - R \cos \sigma)] = \arctan[\rho \sin \sigma/(1 - \rho \cos \sigma)]$. Since $\rho < 1$, α

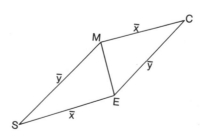

Figure 5.2. Apparent epicyclic motion.

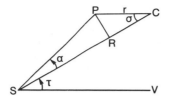

Figure 5.3. Epicyclic angles.

can be expressed as a Fourier series in σ, i.e.,

$$\theta = \tau + \sum_{n=1}^{\infty} \frac{\rho^n}{n} \sin n\sigma.$$

We can, of course, consider this expression as equivalent to

$$\theta = M + \Psi(M - \bar{\omega}) \quad \text{or} \quad M' + \bar{\omega} + \Psi(M').$$

We have seen (p. 114) that in general we can take Ψ in the form

$$\Psi(x) = a_1 \sin M' + a_2 \sin 2M' + \cdots + a_k \sin kM',$$

where the a_i values get progressively smaller. For small eccentricities a_1 is about $2e$, and we can take $\rho = a_1 \cong 2e$. Then the discrepancies between the Keplerian orbit corresponding to Ψ and the above epicyclical model with $\rho = a_1$ will appear only in the higher frequencies. (Notice that we are comparing only angular values. The variation in the distance PS for the epicyclical model used here is twice the variation of PS in the Keplerian orbit, and this can lead to observable differences in the apparent diameter of the sun and planets and in the relative brightness.)

For the apparent motion of the sun relative to the earth, $a_1 = 0.033451$, ρ is 0.033451, and thus the epicyclical expression is

$$\theta = \tau + 0.033451 \sin \sigma + 0.000560 \sin 2\sigma + 0.000012 \sin 3\sigma.$$

The Keplerian second term has value 0.000349 [see Equation (7) above] and the difference 0.000212 radians correspond to 44″. In one day M changes by 0.017203, so that the difference between the two models is about $\frac{1}{80}$ of the change in longitude in a day.

This discrepancy, of course, depends on the eccentricity. If we look at the table on p. 422 of Smart,[24] we see that of classically known planets Mercury has $e = 0.205627$, but Mercury would be hard to observe. For the rest Mars has eccentricity 0.093368, which is seven or eight times as large as that of the earth, and of course, the elliptical character of the orbit of Mars was the first to be established by Kepler.

The "compounding" of epicyclical motions is, of course, very complex, and Copernicus recognized that using the sun as the center simplified the description, since otherwise its motion appeared as an element in the motion of every planet. Actually what Copernicus pointed out was that there was a far more reasonable description of the cosmos than the fixed-earth and earth-centered classical model and in this more reasonable model the larger sun was the center around which the planets moved, including a daily rotating earth. One aspect of the new model was that the previous epicyclical information now indicated approximately the ratio of the orbital major axis of a planet to that of the earth. Thus, the Copernican model was based on two scales. One scale applied to the earth–moon system and could be considered to refer to a reasonably correct estimate of the radius of the earth. The larger scale, of course, involved the orbit of a planet and was relatively correct in terms of the axis of earth's orbit, but the ratio of the two scales was considerably underestimated. Thus, Copernicus considered the ratio of the major axis of the earth's orbit to its radius to be 1142, while it is actually about 23,500. This situation can be described as a reasonable interpretation of the angular information of limited accuracy.

For further discussion of the material in this section, see Neugebauer[20] and Van der Waerden.[29]

5.11. Astronomical Experience in Terms of Accuracy

The development of astronomy is a clear-cut example of interacting intellectual and experimental procedures. In the initial stages the above arithmetical and geometric procedures were adequate to correspond to relatively accurate long-time averages and presumably rather crude angular measurements. Direct observations of angles were based on instruments called "quadrants" or devices that set up isosceles or right triangles so that the angles could be determined by trigonometric tables. Hand-held instruments of this type were probably not more accurate than a tenth of a degree. For angles in a fixed plane, usually the meridian, it was probably possible to set up larger instruments with finer angular resolution, but there may not have been a corresponding increase in accuracy. To obtain a correct order of magnitude estimate of the ratio of the two scales of the Copernican model by direct geometric methods, accuracies of a fraction of a second between cooperating observers at a considerable distance apart would be required.

The accuracy of angular measurements can be inferred to a certain extent from examples quoted in Dreyer. In measuring the radius of the earth Eratosthenes concluded that the difference in latitude between Alexandria and Syrene corresponds to $\frac{1}{50}$ of a circular circumference. Of course, no precision is indicated by this statement, but the corresponding degree expression is $7°\,12'$. The correct value is $7°\,6.7'$, and thus the error is about a tenth of a degree. Ptolemy gave the latitude of Alexandria as $30°58'$, while the true latitude $31°11.7'$, or a difference of two-tenths of a degree. Similarly, Eratosthenes' measurement of the latitude of Syrene was $23°51.3'$ and the true latitude was $23°43.3'$, a difference of $8'$. Tycho Brahe corrects an error of $3'$ in the work of Copernicus relative to the latitude of Ermsland. These are all measurements associated with angles along a meridian and furthermore can be taken near the zenith, where refraction is negligible. Thus, the limit of accuracy was about a tenth of a degree and the associated model of Copernicus is completely appropriate for this experience. The model of Copernicus was not generally accepted for a century after it was proposed, but it was accepted by Kepler, and this proved to be decisive.

The Dark Ages saw a rather complete intellectual deterioration in Western Europe, but the art of instrumentation continued along classical lines in Byzantium and among Arabic astronomers. But from 1300 A.D. on, there was a revival of interest in astronomy in Europe, which, however, was plagued by theoretical inconsistencies and lack of effective observation (see Dreyer[5]). The work of Copernicus is essentially intellectual and his observations are supplements to his development of the model. The Danish astronomer Tycho Brahe realized the importance of systematic observations of the best possible accuracy and set up an observatory with instruments capable of consistent accuracy of the order of a quarter of a minute. These preoptical instruments and the consistent system of observations using them permitted Kepler to correct the epicyclic models into the geometrically sound elliptical orbits and to infer his famous "laws" with their dynamic content that was recognized by Newton. Kepler's analysis uses geometric arguments.

The instruments of Tycho Brahe and his own account of these and his astronomical experience is fascinating (Raeder et al.[23]). He describes the various instruments and the procedures for using them and makes very definite statements about the precision and accuracy of the measurements.

There is one group of instruments for measuring azimuth and elevation or zenith distance. Certain of them consisted of a "quadrant" rotating around a vertical axis. The zenith distance of an object was measured by sighting along the arm of the quadrant. Azimuth measurements were obtained by presetting the azimuth reading and taking the elevation reading at the time the preset azimuth was obtained and noting the time. The quadrants used had radii of from 155 cm to 194 cm. This corresponds to lengths of from 2.7 to 3.386 cm for a degree and one minute corresponds to about half a millimeter. There was a special zigzag scale to permit a fine resolution. Accuracies of $\frac{1}{3}, \frac{1}{4}$, or $\frac{1}{6}$ of a minute were claimed. This must be very close to the limit of visual discernment. See Miczaika and Sinton,[16] pp. 54–55, where it is stated that the eye can separate points as close as one minute of arc.

Another type of instrument for measuring azimuth and elevation had, in place of the quadrant, a combination of rulers that formed an isosceles triangle in which one of the angles corresponded to the zenith distance or the elevation. Copernicus used instruments of this type.

Large "equatorial" instruments were also available in which a circle used as a quadrant was pivoted about an axis parallel to the polar axis. The quadrant reading yielded declension and the amount of rotation yielded right ascension. The latter reading was obtained by presetting the right ascension value, and in order to obtain a reading accurate to within a quarter of an angular minute, the reading had to be taken to within one second in time. This seems to have been the accuracy objective, that is, angles within a quarter of a minute and time to within a second. Four clocks were used in order to obtain the required consistency, and measurements were made independently on three different instruments.

There was also a "zodiacal" instrument in which an interior ring carried a representation of the ecliptic, but this instrument presented difficulties because of lack of balance. Because of this measurements were made on the "equatorial" or fixed system and were converted to longitude and latitude by trigonometry. There was another type of instrument that was used to measure angular differences on the celestial sphere.

For further discussion of the material in this section, see Dreyer,[5] Raeder et al.,[23] and Miczaika and Sinton.[16]

5.12. Optical Instruments and Developments

Thus, before the introduction of the telescope, increasingly sophisticated observations led to the Keplerian orbital formulation and Newton's dynamic description of the solar system. The telescope removed all doubt about the heliocentric character of the solar system, but it also provided greatly increased accuracy to complement the far more precise Newtonian description. Photography provided great versatility for optical instruments and there has also been a tremendous improvement in time measurements.

Telescopes are usually specialized for specific purposes, and angular resolution is not always a prime objective. But for a photographic instrument intended for angular measurements, the relevant aspect is the plate scale, that is, the angular variation that corresponds to a millimeter on the photographic plate. Photography will probably permit a further resolution of about one in a hundred, but other instrumental limitations may negate any such resolution. Miczaika and Sinton[16] state that specialized telescopes have plate scales of 1″ to 2″. Construction problems limit refracting telescopes to a resolution of $4.65/D$ seconds of arc, where D is the diameter of the objective lens in inches. This would correspond to a resolution of a tenth of a second for even reasonably large instruments. For stellar parallaxes Smart[24] (p. 412) indicates a somewhat more refined resolution.

The various ratios in the orbital scale of the solar system are rather precisely determined by Kepler's third law and observations of the orbital periods. This permits one to express various distances in terms of the "astronomical unit," the major axis of the earth's orbit. Thus, if one orbital scale distance can be expressed in terms of the earth's radius, the description of the solar system can be completed. A transit of Venus across the disk of the sun did permit a cooperative parallax measurement of its distance from earth, and the planetoid Eros was also utilized for parallax measurements. Another procedure for determining the size of the solar system involves a precise prediction of satellite motion for a major planet. At different orbital positions this phenomenon has different apparent delays due to the time it takes light to travel the different distances involved. Since the speed of light is known, the distances can be determined. Modern radar has permitted relatively direct and accurate measurements of the distance to the Moon and Venus. Space probes also reveal the fine structure of the dynamic description of the solar system.

Our understanding of the solar system involves a conceptual image including notions of space, i.e., geometry, time and objects subject to certain physical laws. These concepts are scientifically significant because they lead to intellectual mathematical experience that matches actual experience in a very significant manner. It is important to appreciate that both the intellectual image and the actual experience grew by interacting with each other. At every stage the conceptual image provided the format for the experience, and the latter, in turn, reacted on the former. The process is essentially inseparable.

The astronomy of the solar system is an excellent example of how the horizons of human experience have been widened in conjunction with intellectual developments. The modern equivalent is astrophysics with its wide range of observational techniques, including spectroscopy, cosmic ray detection, and visual, radio, and even x-ray telescopy. The corresponding conceptual equivalent involves the modern understanding of atomic and nuclear phenomena and thermodynamic and cosmologic notions. These are expressed in a great range of mathematical concepts. The prediction arithmetic and numerical modeling require the capacity of modern electronic computers. But all this grew in stages that can be precisely determined in history from the sawtooth arithmetic procedure of the Babylonians.

There is, however, one further contrast that must be drawn between the Babylonian arithmetic description of phenomena and certain competing notions. When a merchant at the beginning of a voyage offered sacrifices to Zeus or Poseidon, he acted on the assumption that these great spirits controlled the air and the sea and could be influenced to act in the merchant's favor. Thus, the behavior of the sea and air is explained "animistically." On the other hand if the heavenly bodies move according to a mathematical law that predicts their motion, then clearly no spirit intervenes. Thus, there is a fundamental antithesis between an "animistic" concept of nature and a "mathematical description" of phenomena. A scientific approach always assumes that one is dealing with kinds of experience that have precise logical delimitations and for which relations can be expressed unambiguously in symbolic form. Animistic intervention is inconsistent with such relations.

This antithesis is still completely valid even when the mathematical description involves the notion of probability. The biological theory of evolution involves probability both in the mechanism of inheritance and preferential survival but specifically denies animistic intervention in the evolution of species. Similarly, games of chance are played on the assumption that the outcome is subject to the mathematically described laws of probability. Any animistic intervention is considered cheating.

There has been some confusion in this last regard because of the contrast between the deterministic and probablistic aspects of physics. Actually the latter notions are really complementary rather than antithetical. They both correspond to mathematical descriptions of phenomena and both are antithetical to any animistic description.

For further discussion of the material in this section, see Miczaika and Sinton,[16] Smart,[24] Thackeray,[28] and Whitehead.[31]

For further discussion of the material in this chapter, see Chace,[3] Dijksterhuis,[6] Heath,[10] Moon,[17] Neugebauer,[20] Neugebauer and Satz,[21] and Struik.[26]

Exercises

5.1. Find the formula for the Tartaglia solution of the cubic. Each time a root is taken, a number of possibilities appear so that this procedure apparently yields 18 forms for the solution. Are all 18 forms solutions? Can three distinct solutions be obtained? What are the various fields involved in solving this equation and what are the splitting fields? Can we have complex numbers in the algorithmic procedure even when all the roots of the original equation are real?

5.2. Do the equivalent of the previous exercise for the quartic. One way of obtaining an auxiliary cubic equation for a quartic with roots α_1, α_2, α_3, α_4 is to set up the equation with root $\alpha_1\alpha_2 + \alpha_3\alpha_4$. What are the other roots one should use? How do you set up the equation? Suppose you have solved this cubic. How do you solve the quartic? What is the relation of y to $\alpha_1\alpha_2 + \alpha_3\alpha_4$?

5.3. In the transformations of the one-parameter Lorentz group, c is a constant and the ratio $\beta = v/c$ is considered to be the parameter. Let $T(\beta)$ denote the transformation: $T(\beta)$, $(x, t) \rightarrow (x', t')$, $x' = (x - \beta ct)(1 - \beta^2)^{-1/2}$, $ct' = (ct - \beta x)(1 - \beta^2)^{-1/2}$. Show that $T(\beta_3) = T(\beta_2)T(\beta_1)$ for $\beta_3 = (\beta_1 + \beta_2)/(1 + \beta_1\beta_2)$.

5.4. A finite subgroup of the group of rotations can be associated with a set of unit vectors each of which is the axis of an element of the subgroup. This set of unit vectors is permuted by the elements of the finite subgroup. This permits one to determine all such finite subgroups and in particular those subgroups that leave an infinite lattice invariant. (See Jansen and Boon,[13] page 334.)

5.5. Describe the apparent angular motion of a point on a epicycle to an observer at the center of the fixed circle by a kinematic representation of moving vectors. Compare this procedure with those used by Apollonius as described in Van der Waerden,[29] p. 238–240.

5.6. Consider two spherical coordinate systems, (α, β) and (λ, μ), on a sphere, with α and λ being the longitude variable in each case. One can replace α by $\bar{\alpha} = \alpha + \bar{\alpha}_0$ and λ by $\bar{\lambda}' = \lambda + \bar{\lambda}_0$ so that for the coordinate pairs $(\bar{\alpha}, \beta)$ and $(\bar{\lambda}', \mu)$, $\bar{\alpha}$ and $\bar{\lambda}'$ are measured from a common intersection point Q_0 of the equatorial circles. Let ε be the acute angle between the equatorial circles. We can also replace $\bar{\lambda}'$ and μ by new variables $\bar{\lambda}$ and $\bar{\mu}$, with $\bar{\lambda} = \pm\bar{\lambda}'$ and $\bar{\mu} = \pm\mu$, choosing the appropriate sign so that $\bar{\lambda}$ and $\bar{\alpha}$ have the same sign on adjacent sides of ε and the distance between the positive β and μ poles is ε (Figure 5.4).

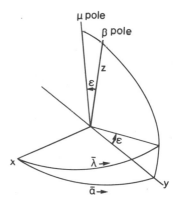

Figure 5.4. Spherical coordinate systems.

Show that

$$\tan \bar{\lambda} = \cos \varepsilon \tan \bar{\alpha} - \sin \varepsilon \sec \bar{\alpha} \tan \beta$$
$$\sin \bar{\mu} = \sin \varepsilon \cos \beta \sin \bar{\alpha} + \cos \varepsilon \sin \beta.$$

[*Hint*: Take a rectangular coordinate system x, y, z corresponding to the $(\bar{\alpha}, \beta)$ coordinates for which $x = \cos \bar{\alpha} \cos \beta$, $y = \sin \bar{\alpha} \cos \beta$, $z = \sin \beta$. Rotating an amount ε about the x axis will yield the corresponding Cartesian system for $(\bar{\lambda}, \bar{\mu})$.]

Apply this to the equatorial and zodiacal coordinates on the celestial sphere. Express the relation between azimuth and elevation and the equatorial coordinates of a star using siderial time.

5.7. Given

$$\frac{dv/dt}{(1 + e \cos v)^2} = \frac{n}{(1 - e^2)^{3/2}},$$

let $e^* = e/[1 + (1 - e^2)^{1/2}]$ and $t = \bar{t}$ for $v = 0$. Show that $n(t - \bar{t}) = v + \sum_{n=1}^{\infty} (2(-1)^n e^{*n} \times [(1 - e^2)^{1/2} + 1/n] \sin nv$.

R

1.
2. G
   G
3. Ch
   Pap
4. Dau
5. Drey
   Inc., N
6. Dijkste
   (1956).
7. Einstein,
   (1952).
8. Forder, H.
   (1958).
9. Forder, H. G
10. Heath, T. L.,
    Kingdom (191
11. Heath, T. L., T
    New York (1956
12. Hilbert, D. (1956
13. Jansen, L. and Bod
    Amsterdam/John V
14. Lines, L., Solid Geon
15. Manning, H. P., Intro
    (1963).
16. Miczaika, G. R., and
    Cambridge, Massachus
17. Moon, P., The Abacus,
18. Morrow, G. R., Proclus:
    University Press, Princeto
19. Murray, F. J., Mathematic
20. Neugebauer, O., The Exact
    York (1969).

21. Neugebauer, O., and Satz, A., *Mathematical Cuneiform Texts*, American Oriental Series. Vol. 29 (1945); published jointly by the American Oriental Society and the American Schools of Oriental Research, New Haven, Connecticut.

22. Peet, T. E., *Mathematics in ancient Egypt*, Bulletin of the John Rylands Library, **15**(2) (July 1931), Manchester University Press, Manchester, England.

23. Raeder, H., Stromgen, E., and Stromgen, B., *Tycho Brahe's Description of His Instruments and Scientific Work*, Royal Danish Academy of Sciences and Letters, Copenhagen (1946).

24. Smart, S. M., *Text-Book on Spherical Astronomy* (5th edition), Cambridge University Press, Cambridge, United Kingdom (1962).

25. Smith, D. E., *A Source Book in Mathematics* McGraw-Hill Book Co., New York (1929).

26. Struik, D. J., *A Concise History of Mathematics*, Dover Publications, Inc., New York (1948).

27. Struve, W. W., Mathematische Papyrus der staetlichen Museum der Schoenen Kunste, *Quellen und Studien zur Geschichte der Mathematik*, Vol. 1, Part A, J. Springer Berlin (1930).

28. Thackeray, A. D., *Astronomical Spectroscopy*, The MacMillan Company, New York (1961).

29. Van der Waerden, B. L., *Science Awakening*, P. Noordhoff Ltd., Gröningen, Holland (1954).

30. Ver Eeke, Paul, *Diophante d'Alexandrie*, Albert Blanshard, Paris (1959).

31. Whitehead, T. N., *Instruments and Accurate Mechanisms* (reprint), Dover Publications, Inc., New York (1954).

re
x.

$X_1 =$
funct
values
formul
one inv
5.11
$X(\alpha) = (si$
inversion

then for $\alpha \neq \mu$

$X(\alpha) \cdot$
and $X(\alpha) \cdot X(\alpha) =$
vectors? How ca
5.12. Show

5.13. On the ba
tions of the earth with
find the points of maxi

6

Natural Philosophy

6.1. Analysis

During the Renaissance there was a tremendous European interest in mathematics and it produced a more sophisticated and effective algebra. This algebra was combined with various geometric procedures and other concepts to yield the methods of analysis. In classical mathematics quantitative methods were applicable only to "numbers," i.e., natural or mixed, and a limited range of geometric magnitudes. The new analysis represented an extension of quantitative procedures to a much larger domain of experience including, kinetics, dynamics, the properties of matter, and to a far more general "analytic" geometry. This was the critical intellectual achievement that produced the modern exact sciences.

The detailed history of this development is available in such references as Klein,[13] Boyer,[4] Smith,[19] and Robinson.[18] Our immediate interest is in the rich variety of intellectual elements that were part of this development and the reasons they were ultimately transformed into the modern form of mathematics. There were at least two distinct conceptual forms for the calculus, and both were different from our present-day mathematics. It is interesting to observe that a good deal of the terminology and symbolism of these early forms of the calculus has been retained, but in order to fit such a term or symbol into our present formulation, it is redefined in a way that seems completely different from the obvious interpretation of the term or symbol.

The European development of algebra had two complementary aspects. One of these was a progressive change in attitude accompanied by an evolution in notation so that, finally, the manipulation of equations was considered to have logical significance equivalent to the arguments of geometry. Thus, one could set down an equation and by justified manipulations obtain necessary characteristics of an unknown, or similarly, one could start with an expression for a number and establish desired properties. The other aspect of this algebra was an increase in capability to include the solution of the cubic and quartic equations and various formulas involving a general n that could be established inductively. For example, the sums of powers of integers are given by

$$1+2+\cdots+n=n(n+1)/2$$
$$1^2+2^2+\cdots+n^2=n(n+1)(2n+1)/6.$$

These formulas readily lend themselves to obtaining areas by classical "methods of exhaustion." Thus, the area under the curve $y=x^2$ between $x=0$ and $x=1$ can be boxed in between two sets of rectangles with areas

$$A'=\sum_{j=1}^{n}\frac{(j-1)^2}{n^2\cdot n}\qquad\text{and}\qquad A''=\sum_{j=1}^{n}\frac{j^2}{n^3},$$

respectively. The above formula yields that $A'=\frac{1}{3}-1/2n+1/6n^2$ and $A''=\frac{1}{3}+1/2n+1/6n^2$. Since we can take n arbitrarily large, A must be $\frac{1}{3}$. For integral n, the area under $y=x^n$ between $x=0$ and $x=a$ can be shown to be $a^{n+1}/(n+1)$.

Another expansion of quantitative procedure was represented by the principle or axiom of Cavalieri. In the case of volumes this principle would state that two volumes are equal if they have equal bases and if the cross sections parallel to the base at equal distances from the base are equal. Thus, if the cross section of one solid is the sum of the cross sections of two other solids, the volume of the first is the sum of the volumes of the other two. As in many other developments of a similar nature at this time, this principle of Cavalieri had a wide range of practical valid application. This range certainly exceeded the classical solids, but there was no conceptual framework in which it could be formulated so that its exact domain of validity could be established. The "proof" given by Cavalieri appears to be a kind of limiting process that we would consider as valid only under a considerable number of additional assumptions.

This Archimedean comparison of cross sections could be applied to both moments and linear motion. The moment of the area of the triangle under $y=x$ and between $x=0$ and $x=1$ considered about the y axis is equal to the volume of a pyramid with base perpendicular to the x axis and with cross section of area x^2. This is readily generalized so that moments of areas between curves with positive ordinates and the x axis and between specified x limits and around the y axis can be equated to volumes.

For linear motion it was known that if one takes time along an axis, say, the x axis, and at each abscissa erects an ordinate equal to the speed, then the distance covered in the time between t_1 and t_2 equals the corresponding area.

It was also known that if one takes the ordinate equal to the distance covered from a fixed time t_0, then the slope of the tangent to this curve was equal to the speed. In a sense this was a form of the fundamental theorem of the calculus that was of great practical significance.

For further discussion of the material in this section, see Boyer,[4] Klein,[13] Robinson,[18] and Smith.[19]

6.2. The Calculus

During the seventeenth century, procedures for finding tangents by algebraic methods appeared in many forms. The "symptom" relations were translated into algebraic forms that readily permitted these manipulations. For example, let us find the slope of the tangent to $x^2 + 2y^2 = 3$ at $(1, 1)$. Consider a point $(x', y') = (1 + h, 1 + k)$ on the ellipse. Thus, $2h + h^2 + 4k + 2k^2 = 0$. Dividing by h yields

$$(2 + h) + \frac{k}{h}(4 + 2k) = 0$$

Now if we consider the point (x', y') as moving along the ellipse toward $(1, 1)$, then the ratio k/h is also a changing quantity, i.e., a "variable" that approaches $-\frac{1}{2}$ as (x', y') moves to $(1, 1)$. From the practical point of view all one really had to do was set h and k equal to zero in the above equation, ignoring the fact that they occur in the ratio that one considers a new quantity.

Clearly, with the idea of motion one has a considerable number of new concepts, for example, "variable," "limit" as in the above, and the notion of "function," which is a relation between variables analogous to the notion of ratio as a relation between magnitudes. Newton's notion of limit can be summarized by Lemma 1 of Book One of the *Principia*[6]: "Quantities and ratios of quantities which in any finite time converge continually to equality and before the end of that time approach nearer the one to the other than any given difference become ultimately equal." In the next lemma, the "boxing" process for an area is thought of as being carried out in time with the sides of the rectangles "diminishing *in infinitum*." Similarly the ratio of chord to arc on a curve is considered as the end points "approach one another." He insists in the *Scholium* at the end of this sequence of lemmas that he is not concerned with "indivisibles," i.e., infinitely small quantities, but with the "ultimate ratio" of ratios of finite quantities. Of course, when the two magnitudes in a ratio are zero there is no ratio, but the algebraic process finds one. To bridge this gap some concept of "continuity" or simply the existence of the limit is required. For a moving "variable" this notion of continuity is intuitive and implicit.

Newton incorporated these notions of limits into the calculus in the form of the theory of "fluxions" and "fluents" of "variables." In present-day terminology this would be the theory of derivatives or indefinite integrals of functions of time. For these the available algebra, including infinite series, permitted a wide range of application. The limiting processes were to a considerable extent intuitive, with a free exchange of the order in limiting processes. Consider, for example, the differentiation of x^n for n nonintegral. Let the increment of x be denoted by O. Then Newton had established that

$$(x+O)^n = x^n + nx^{n-1}O + n(n-1)x^{n-2}O^2/2 + \cdots.$$

This formula had been inferred by Newton by analogy with the case of n integral. For $n = \frac{1}{2}$ this can also be justified by multiplying the series by itself term by term. The ratio of the change of x^n to O, the change in x, is that of $nx^{n-1} + n(n-1)x^{n-2}O/2 + \cdots$ to 1 and the limit is taken by setting O to zero. This can be considered to be either an exchange of order for the limits or an assumption of continuity.

There was of course a considerable amount of criticism of Newton's calculus and that of Leibnitz. This did not inhibit mathematicians of the succeeding era from obtaining and using results such as

$$\tfrac{1}{4} = 1 - 2 + 3 - 4 + \cdots,$$

which is a consequence of an interchange of limits. In general when rigorous standards were established such illegitimate offspring were viewed with disdain. However, later, methods of handling precisely such relations were established and the results were justified.

It has not been unusual in mathematics to have the symbolism and formal manipulation of a subject extended beyond the original conceptual development. The consequent effort to increase the conceptual framework has often resulted in important mathematical advances. The notion of higher dimensions in geometry, of projective geometry with its points at infinity, algebraic ideal theory, and the theory of distributions are examples.

Thus, while logical criticism could be directed at the new analysis, its most important characteristic was the expansion of mathematics, which permitted one to express the new scientific outlook. In Newton's *Principia* this was done for the solar system as far as it was known. Kepler's laws were shown to be equivalent to Newtonian mechanics and universal gravitation, and a tremendous range of phenomena such as the tides and precession of the equinox was explained. The effect of this could only be described as an intellectual revolution. There were also practical effects of matching importance. The theory of dynamics was essential for the development of industrial machinery and the geometric aspect of analysis was adequate to determine, for example, the desirable shape of gear teeth.

The student will find the history of the development of the concepts of mathematics fascinating as available in the references. There was, of course, a rival development of the calculus due to Leibnitz that involved different notions. Thus, instead of a figure being "generated" by a cross section of one less dimension, the figure was considered to be the sum of an "infinite number" of figures of the same dimension but with "infinitesimal height." Algebra was used on "differentials" to obtain "differential coefficients," i.e., either derivatives or partial derivatives. But just as in Newtonian calculus much of the power of the new analysis was based on the fundamental theorem of the calculus, i.e., on antidifferentiation.

While the Leibnitz manipulations were essentially the same as those of Newton, a rather vague "principle of continuity" was supposed to be applicable to the ratio of differentials. This is discussed in the last chapter of Robinson.[18] It is, of course, the Leibnitz notation that has survived, but this is now accompanied by a full kit of explanations and definitions so that a Leibnitz manipulation can be replaced by a modern mathematical procedure. On the other hand the "nonstandard analysis" described in Robinson's book gives an "extension" of the real number system by set theoretic procedures in the modern manner that permits one to interpret directly the language of Leibnitz with "infinitesimally small" and "infinite" objects. However, a satisfactory interpretation is not facile.

For further discussion of the material in this section, see Boyer,[4] Robinson,[18] and Smith.[19]

6.3. The Transformation of Mathematics

The intellectual revolution consequent on the Newtonian scientific advances produced a philosophical approach that married mathematics and experimentation. There was absolute truth accessible to the human mind and expressible in mathematical form, but the choice of the mathematical form had to be verified by experiment on the mathematical consequences. Realms of experience were governed by mathematically expressed laws, and there were complete intellectual satisfaction and important practical benefits in deriving mathematical relations and verifying them by experience.

This philosophical viewpoint could be rather naturally associated with the peripatetic (i.e., Aristotelian) notion that mathematical concepts arose from experience by abstraction. But the nature of magnitudes available from experience had been given by Euclid and Archimedes and did not include infinitesimal quantities or infinite magnitudes. Infinite aggregates were in a somewhat different category. Thus, there were unsatisfactory aspects to the calculus of Leibnitz and in a somewhat different way to that of Newton.

The concern for logical consistency ultimately led to a complete recasting of mathematics. But one major aspect of this was the determination of what was logically satisfactory, a motif that became dominant in the nineteenth century. But in the eighteenth century and even into the nineteenth, the major motivation for the development of mathematics was to increase capability as part of the maturing of the exact sciences.

Logical arguments dealt with equations rather than geometric figures. Newton's *Principia* was an extension of geometry, but the emphasis of succeeding mathematicians was on "analytic" notions such as equations and functions, derivatives, integrals, and series. Both Newton and Leibnitz used antidifferentiation to replace "summation" approaches to integration. But this in turn was generalized to the use of differential equations to solve problems as the central procedure of analysis.

Consequently, mathematics dealt mainly with the numerical value associated with a magnitude. If one chooses a unit length and fixed origin of a Euclidean line, then the notion of ratio can be equated to that of "point" on this line. This yields a concept of "real number," including negatives. Similarly the points of the plane can be associated with complex numbers (see Smith[19]). It was these concepts of real and complex number that were used to express the analytic form of many new concepts, e.g., surfaces, multidimensional analysis, differential equations.

The solution of differential equations often involved infinite series. The formal manipulation of "infinite sums" and "infinite products" went through a period of great flexibility and questionable logical rigor. But by the beginning of the nineteenth century, the requirement of convergence led ultimately to the modern formulations.

The notion of function and the associated notion of limit passed through numerous phases. We have indicated the idea of a function as a relation of moving "variables" and the corresponding "approach to a limit." The Taylor series for the basic transcendental functions and the algebraic functions were available and were clearly effective in computation. This suggested the definition of a function as a formal Taylor series and, correspondingly, the use of infinitesimals of various orders in the limiting processes. The physical notion of "motion" is thus eliminated in favor of computation. A "variable" is then a symbol as in elementary algebra, and the functional relation is operational, i.e., a power series. The formal Taylor series concept ran into difficulties with the appearance of the more general possibility of a Fourier series. Analysis now dealt exclusively with real numbers rather than with the ratio of magnitudes, and notions such as continuity and limit were expressible in terms of inequalities. By means of Fourier series one could construct continuous functions that were not differentiable. It became evident that only the abstract correspondence notion or the equivalent set-theoretic notion of the graph provided the appropriate generality for the function concept. Also, these did provide the possibility for an exact formulation of numerical procedures. The use of inequalities is readily translated into set-theoretic ideas. The latter had many inventive and useful aspects. Thus, the notion of compactness led to a satisfactory discussion of the properties of continuous functions. With the introduction of the constructive possibilities for infinite sets, this new form of analysis was stabilized and received intensive development.

The precision and inventiveness of modern analysis was highly significant for further scientific development. But the direct practical use of the new analysis was far clumsier than the earlier infinitesimal methods. This is evident in any calculus course. The older procedures were reformulated as theorems in the new mathematics using the older notation. Thus, discussions may still be expressed in the earlier language but have a modern significance in terms of numerical procedures.

Thus, mathematics was transformed into a purely numerical subject by eliminating all intuitive notions of space and time. In fact the tables were now turned. Notions of space and time could be described precisely in a numerical fashion. But the conceptual basis of mathematics was irrevocably based on the concept of infinite sets. One can, of course, claim that as far as scientific applications are concerned only the symbolic formulation is essential and that really the conceptual basis plays only a heuristic role.

The discovery of non-Euclidean geometry indicated that mathematics did not necessarily arise from some "true" picture of the universe but was essentially inventive, and this was confirmed by the nature of this new logical formulation of mathematics. The mathematical theory of a science does not have absolute validity but is justified by its agreement with experience, specifically in regard to probing experiments. Mathematics is a conceptual form for expressing experience patterns or in modern jargon a "language." It does not have factual content and the question of "truth" is irrelevant. The existence of a form of mathematics is simply a matter of "freedom from contradiction," i.e., just the capability of a satisfactory logical discussion.

For further discussion of the material in this section, see Boyer.[4]

6.4. The Method of Fluxions

Let us begin by quoting Bishop Berkeley in his famous book of criticism *The Analyst. A Discussion Addressed to an Infidel Mathematician* (1734) (see Smith,[19] p. 628):

> The Method of Fluxions is the general key, by help whereof the modern mathematician unlocks the secrets of geometry and consequently of Nature. And as it is that which hath enabled them so remarkably to outgo the ancients in discovering theorems and solving problems, the exercise and application thereof is become the main if not the sole employment of all those who in this age pass for profound geometers.

The latent sarcasm surfaces at the end of this brief quotation, but his description of the mathematics of the eighteenth century is quite apt. In what way then did the calculus permit the "profound geometers" "to outgo the ancients"? The answer is the following ways: (a) They reasoned with equations instead of or in addition to diagrams. (b) They had analytic tools to treat far more general geometric objects. (c) They dealt with new magnitudes of broad practical significance.

The basic task of geometry had been to structure the quantitative aspects of geometric magnitudes. The calculus permitted the introduction of many new magnitudes to expand the domains of natural philosophy that were subject to quantitative control. On the most elementary level, these magnitudes dealt with the notions of particle movement, but one also had such concepts as that of a field of force or field of fluid flow and ultimately a conceptual framework for the behavior in time of a substance distributed in space.

The "Method of Fluxions" contains a conceptual structure that is highly significant for modern applied mathematics, and we will discuss it without emphasis on the historical development. One can consider geometry as having an analytic formulation, but the major adjustment was that one considered curves and surfaces, which were associated with the elementary notions of lines and planes only in an "infinitesimal" sense. This meant that the final result of a discussion had to be obtained by integration, that is, by solving a system of differential equations.

In classical geometry the geometric definition of a locus led to equations such as the "symptom." But one can use operational equations on the coordinates of a point to specify the geometric loci, greatly expanding the curves and surfaces that can be considered. Alternatively one can use the parametric form based on the notion of function to specify curves or surfaces. A curve in three-space is given by three equations, $x^i = f^i(t)$, $i = 1, 2, 3$, defined for a suitable range of t. A surface is given by $x^i = f^i(r, s)$, where r and s vary over an appropriate region in the plane. To apply the desired analysis differentiability conditions are required, and these do yield a notion of dimensionality. A more modern notion of manifold uses a number of such parametric representations of a surface "patched" together in a suitable way.

If the $f^i(t)$ that specify a curve have continuous first derivatives, there is a natural parameter, the distance s along the curve from a fixed point. If the

$f^i(t)$ have second derivatives and the parameter t is considered to be the time, then the curve can be considered to be the path of a particle, and one can specify the vector notions of velocity and acceleration and the scalar concept of kinetic energy.

Newtonian dynamics utilizes two complementary notions of a moving particle and a field of force. A field of force involves a region, at each point of which a vector is specified. Both the motion of the particle and the force vector are to be expressed in a Cartesian coordinate system for which Newton's third law is valid. This means that there is a property of the particle, such as mass or electric charge, that couples the force and the acceleration. If one has found one such coordinate system, any other that is in uniform translation motion relative to the first will also do.

In the case of gravitational attraction between N particles, the position of the particles determines the field of force. Thus, each of the $3N$ Cartesian coordinates, $y^1, ..., y^{3N}$, of the particles satisfies a differential equation

$$m\ddot{y}^k = F_k(y^1, ..., y^{3N}).$$

This system of second-order differential equations determines the motion of the N particles, provided the positions and velocities are given at an instant of time. In the case of the solar system numerical solutions of the equations are available to within the observational accuracy. The initial success of Newton's theory was its application to astronomical problems.

The notion of work is associated with that of a force acting through a distance. If a particle moves through a field of force, the field will do work on the particle that can be expressed as an integral. If the particle is displaced the vector amount (dx^1, dx^2, dx^3) due to a change of parameter dt, then the work done by the field $\{F^1, F^2, F^3\}$ is $dW = F^1 dx^1 + F^2 dx^2 + F^3 dx^3$, and thus the total work done by the field as the particle transverses the path $\mathfrak{C}$ is

$$W = \int_a^b \left(F^\alpha \cdot \frac{dx^\alpha}{dt} \right) dt = \int_{\mathfrak{C}} F^\alpha \, dx^\alpha.$$

This expression is independent of the choice of the parameter t, which need not be the time. If the field of force does not depend on the time, this integral depends only on the curve $\mathfrak{C}$ and is termed a "line integral." The work expression has been obtained on the basis of an analysis using the inner product of an infinitesimal displacement with the field-force vector.

6.5. The Behavior of Substance in the Eulerian Formulation

Euler proposed a procedure for describing the behavior of a substance that is particularly effective in the case of a moving fluid, but it also has wider applications. One considers a region of space with a coordinate system x^1, x^2, x^3 and an interval of time. At any instant in time, t, the substance occupies part or all of this region, so that its behavior can be described by numerical procedures on functions of x^1, x^2, x^3, and t. Thus, for example, we can obtain an approximate description by forming, for each of a discrete set of values of t, a spatial partition with subdivisions of small diameter. Usually one can assume that any discontinuities in the behavior of the substance occur on the boundary of subdivisions, so that one can treat the substance within a subdivision as a particle with properties associated with some point in the substance. The theoretical particle has a mass similar to the mass of the substance in the subdivision and has a velocity and acceleration or other relevant properties. The values of these quantities are given by certain functions of x^1, x^2, x^3, and t, and one can indicate the limitations on the nature and behavior of the substance so that partitions of this type can be effectively used to approximate the behavior of the substance. Such approximations are still useful even if one insists on the atomic structure of matter, since in many cases one can have a very fine partition of the material to justify the particle approximation, and yet this partition is coarse relative to the atomic structure.

It is practical then to consider the substance as being described by functions of space and time in which the ultimate atomic fine structure is ignored. Quantities having practical experimental significance for an instant of time, t, arise by integration over a volume or in certain cases over a surface. The properties of the substance that one associates with the integrands— that is, with functions of points in the substance—are termed "intensive" and the quantities obtained by integration are called "extensive." Thus, the density, $\rho(x^1, x^2, x^3, t)$, is an intensive function that yields the mass in a region, $\mathfrak{A}$, by integration

$$M = \iiint_{\mathfrak{A}} \rho \, dV.$$

The velocity of a point in the substance is a vector, $v(x^1, x^2, x^3, t)$, with components v^1, v^2, v^3, which yields the momentum, P, in a region, $\mathfrak{A}$, at an instant of time by

$$P = \iiint_{\mathfrak{A}} \rho v \, dV$$

and the kinetic energy

$$T = \frac{1}{2} \iiint_{\mathfrak{A}} \rho v \cdot v \, dV.$$

For a homogeneous substance intensive quantities do not depend on the amount of substance, while extensive quantities are proportional to the amount.

At an instant of time, t, an intensive quantity $u(x^1, x^2, x^3, t)$ can be considered to denote a property of the substance associated with a small subdivision of the spatial partition. We suppose that this amount of the substance is identifiable, at least for a small time, dt. To determine how this local property, u, of the substance changes in time, dt, we must consider the motion of the substance. Thus, the change

$$du = u(x^1 + v^1\, dt, x^2 + v^2\, dt, x^3 + v^3\, dt, t + dt) - u(x^1, x^2, x^3, t)$$

$$= \left(\frac{\partial u}{\partial x^1} v^1 + \frac{\partial u}{\partial x^2} v^2 + \frac{\partial u}{\partial x^3} v^3 + \frac{\partial u}{\partial t} \right) dt.$$

Hence we can define the "intrinsic rate of change" of u:

$$\frac{Du}{Dt} = \frac{\partial u}{\partial x^1} v^1 + \frac{\partial u}{\partial x^2} v^2 + \frac{\partial u}{\partial x^3} v^3 + \frac{\partial u}{\partial t}.$$

This gives the rate of change of a property associated with a moving part of the substance.

Surface integrals are also used to describe the behavior of substances. Consider a surface or part of a surface given by a set of equations, $x^i = x^i(r, s)$ for $0 \leqslant r \leqslant 1, 0 \leqslant s \leqslant 1$. We partition the unit r interval and the unit s interval, $0 = r_0, r_1, \ldots, r_n = 1$, $0 = s_0, s_1, \ldots, s_n = 1$, and consequently the surface into quadrilaterals, σ_{ij}, for which $r_{i-1} \leqslant r \leqslant r_i$, $s_{j-1} \leqslant s \leqslant s_j$. Let P'_{ij} be a point in σ_{ij} with corresponding r'_i and s'_j. Let

$$R = R(r'_i, s'_j) = \left(\frac{\partial x^1}{\partial r}(r'_i, s'_j), \frac{\partial x^2}{\partial r}(r'_i, s'_j), \frac{\partial x^3}{\partial r}(r'_i, s'_j) \right)$$

$$S = S(r'_i, s'_j) = \left(\frac{\partial x^1}{\partial s}(r'_i, s'_j), \frac{\partial x^2}{\partial s}(r'_i, s'_j), \frac{\partial x^3}{\partial s}(r'_i, s'_j) \right).$$

The vectors $R\,\Delta r$ and $S\,\Delta s$ lie in the tangent plane to the surface at P'_{ij} and determine a planar quadrilateral that approximates σ_{ij} when the tangent vectors are appropriately located. The vector $(R \times S)\,\Delta r\,\Delta s$ is normal to the surface at P'_{ij} and has the area of the planar quadrilateral as its length. Thus, $(R \times S)\Delta r\,\Delta s = n\,\Delta A$, where n is a unit vector in the normal direction and ΔA is the quadrilateral area. This construction permits one to approximate various quantities by appropriate sums. We will assume that n always appears on the same side of the surface. This side will be the "upper side."

Let us consider the volume of substance that passes from the lower side of the surface to the upper in a short period of time dt. Let v' denote the velocity of the substance at P'_{ij} so that, in the time dt, the substance at P'_{ij} will move to the point $P'_{ij}+v'\,dt$. We consider the planar quadrilateral approximating σ_{ij} as also moving in this way so that that substance that passes through σ_{ij} in time dt occupies a parallelepiped with the σ_{ij} quadrilateral as a base and $v'\,dt$ as slant edge. This parallelepiped has altitude $(v'\,dt)\cdot n$ and volume $dV = v'\cdot n\,dA\,dt = v'\cdot(R\times S)dr\,ds\,dt$. This yields

$$\frac{dV}{dt}=\iint v\cdot(R\times S)\,dr\,ds.$$

One also has that the rate at which the mass of the substance passes through the surface is

$$\frac{dM}{dt}=\iint \rho v\cdot(R\times S)\,dr\,ds.$$

For either of these integrals we can proceed in the following way:

$$\frac{dV}{dt}=\iint v\cdot(R\times S)\,dr\,ds$$

$$=\iint v^1\frac{\partial(x^2,x^3)}{\partial(r,s)}\,dr\,ds+\iint v^2\frac{\partial(x^3,x^1)}{\partial(r,s)}\,dr\,ds+\iint v^3\frac{\partial(x^1,x^2)}{\partial(r,s)}\,dr\,ds.$$

We assume that

$$J_{23}\equiv\frac{\partial(x^2,x^3)}{\partial(r,s)}=\frac{\partial x^2}{\partial r}\frac{\partial x^3}{\partial s}-\frac{\partial x^3}{\partial r}\frac{\partial x^2}{\partial s}$$

is continuous and that the surface can be divided into a finite number of regions on which either J_{23} is zero or such that the relations $x^2=x^2(r,s)$, $x^3=x^3(r,s)$ can be inverted and r and s expressed as functions of x^2 and x^3. In the latter case x^1 will also be a function of x^2 and x^3 and we have

$$\iint v^1\frac{\partial(x^2,x^3)}{\partial(r,s)}dr\,ds=\sum\iint v^1(x^1(x^2,x^3),x^2,x^3)\,dx^2\,dx^3$$

$$=(\text{say})\iint v^1\,dx^2\,dx^3.$$

By a proper interpretation of the integrals, we may write

$$\frac{dV}{dt}=\iint(v^1\,dx^2\,dx^3+v^2\,dx^3\,dx^1+v^3\,dx^1\,dx^2),$$

and we have a similar expression for dM/dt.

The integral

$$\int f^\alpha(x^1,x^2,x^3)\,dx^\alpha=\int f^\alpha\frac{dx^\alpha}{dt}\,dt$$

is unchanged if the parameter in the parametric equations for $\mathfrak{C}$ is changed.

Replacing t by another parameter, t', requires that t be a function $h(t')$ of t', and the chain rule for differentiation shows that the integral is unchanged. Similarly if one replaced the pair of parameters, r, s, for a surface by another pair, r', s', then r and s are functions of r' and s' and one can show that

$$\frac{\partial(x^2, x^3)}{\partial(r', s')} = \frac{\partial(x^2, x^3)}{\partial(r, s)} \frac{\partial(r, s)}{\partial(r', s')}.$$

The rule for changing variables in a double integral will now show that the surface integral is not affected by a different parametrization.

The three expressions $f^\alpha \, dx^\alpha$, $v^1 dx^2 dx^3 + v^2 dx^3 dx^1 + v^3 dx^1 dx^2$, and $dx^1 dx^2 dx^3$ that appear under the integral signs are examples of "differential forms." The procedures for handling such forms and the associated integrals are well developed (see Flanders[18]. A product such as $dx^2 dx^3$ or $dx^1 dx^2 dx^3$, which occurs in these forms, is not the usual product of differentials but should be interpreted as

$$dx^2 \, dx^3 = \frac{\partial(x^2, x^3)}{\partial(r, s)} \, dr \, ds,$$

where r and s are independent variables. Consequently we have such properties as $dx^3 \, dx^2 = -dx^2 \, dx^3$ and $dx^1 \, dx^1 = 0$.

6.6. The Generalized Stokes' Theorem

There is a very important relation involving differential forms that is illustrated by Gauss' theorem. Suppose $\mathfrak{B}$ is a surface bounding a three-dimensional region $\mathfrak{A}$. Then

$$\iint_{\mathfrak{B}} (f^1 \, dx^2 \, dx^3 + f^2 \, dx^3 \, dx^1 + f^3 \, dx^1 \, dx^2)$$

$$= \iiint_{\mathfrak{A}} \left(\frac{\partial f^1}{\partial x^1} + \frac{\partial f^2}{\partial x^2} + \frac{\partial f^3}{\partial x^3} \right) dx^1 \, dx^2 \, dx^3 = \iiint_{\mathfrak{A}} (\text{div } f) dx^1 \, dx^2 \, dx^3.$$

This permits one to associate an intensive quantity, div f, with a vector field. For example, suppose that the region $\mathfrak{A}$ is included in the part of space occupied by a certain substance at time t and has boundary $\mathfrak{B}$. The substance in $\mathfrak{A}$ will change its volume at a rate given by

$$\frac{dV}{dt} = \iint_{\mathfrak{B}} (v^1 \, dx^2 \, dx^3 + v^2 \, dx^3 \, dx^1 + v^3 \, dx^1 \, dx^2)$$

$$= \iiint_{\mathfrak{A}} \left(\frac{\partial v^1}{\partial x^1} + \frac{\partial v^2}{\partial x^2} + \frac{\partial v^3}{\partial x^3} \right) dx^1 \, dx^2 \, dx^3,$$

where $v^i = v^i(x^1, x^2, x^3, t)$ is the ith component of the velocity vector for the substance. Clearly we can regard div v as a growth rate or local coefficient of expansion for $dV = dx^1\, dx^2\, dx^3$, i.e.,

$$\frac{d}{dt}(dV) = \left(\frac{\partial v^1}{\partial x^1} + \frac{\partial v^2}{\partial x^2} + \frac{\partial v^3}{\partial x^3}\right) dV.$$

An incompressible substance must satisfy div $v = 0$. In vector notation, one can write Gauss' theorem as

$$\iint_{\mathfrak{B}} f \cdot n\, dA = \iiint_{\mathfrak{A}} \operatorname{div} f\, dV.$$

Suppose now that $\mathfrak{A}$ is a fixed region in space occupied by a substance. The mass M in the region $\mathfrak{A}$ is given by $M = \iiint \rho\, dV$, and the decrease in mass in the region is

$$-\frac{dM}{dt} = -\iiint_{\mathfrak{B}} \frac{\partial \rho}{\partial t}\, dV.$$

But this is also the rate at which material is leaving $\mathfrak{A}$ through the boundary $\mathfrak{B}$ of $\mathfrak{A}$:

$$-\iiint_{\mathfrak{A}} \frac{\partial \rho}{\partial t}\, dV = \iint_{\mathfrak{B}} (\rho v^1\, dx^2\, dx^3 + \rho v^2\, dx^3\, dx^1 + \rho v^3\, dx^1\, dx^2)$$

$$= \iiint_{\mathfrak{A}} \left[\frac{\partial(\rho v^1)}{\partial x^1} + \frac{\partial(\rho v^2)}{\partial x^2} + \frac{\partial(\rho v^3)}{\partial x^3}\right] dV.$$

Since this holds for any such region $\mathfrak{A}$, one must have

$$\frac{\partial \rho}{\partial t} + \frac{\partial(\rho v^1)}{\partial x^1} + \frac{\partial(\rho v^2)}{\partial x^2} + \frac{\partial(\rho v^3)}{\partial x^3} = 0.$$

Thus, the density of a substance satisfies this equation, which is termed "the equation of continuity."

A similar result is Stokes' theorem. Let $\mathfrak{B}$ be a part of a surface bounded by a curve $\mathfrak{C}$. Then

$$\int_{\mathfrak{C}} f^1\, dx^1 + f^2\, dx^2 + f^3\, dx^3 = \int_{\mathfrak{C}} f \cdot dx = \iint_{\mathfrak{B}} \left(\frac{\partial f^3}{\partial x^2} - \frac{\partial f^2}{\partial x^3}\right) dx^2\, dx^3$$

$$+ \left(\frac{\partial f^1}{\partial x^3} - \frac{\partial f^3}{\partial x^1}\right) dx^3\, dx^1 + \left(\frac{\partial f^2}{\partial x^1} - \frac{\partial f^1}{\partial x^2}\right) dx^1\, dx^2$$

$$= \iint_{\mathfrak{B}} (\operatorname{curl} f) \cdot n\, dA.$$

Stokes' and Gauss' theorems are three-dimensional examples of a more general result. Suppose we have an integral applicable to a k-dimensional subspace in an n-dimensional space. Then

$$\iiint_{\mathfrak{B}} \Sigma f^{i_1,\ldots,i_k} dx^{i_1} \ldots dx^{i_k}$$

$$= \iiiint_{\mathfrak{A}} \Sigma \left[\Sigma (-1)^{j-1} \frac{\partial f^{i_1,\ldots,[ij],\ldots,i_{k+1}}}{\partial x^{ij}} \right] dx^{i_1} \ldots dx^{i_{k+1}},$$

where $\mathfrak{B}$ is the k-dimensional boundary of the $(k+1)$-dimensional region $\mathfrak{A}$. The subscripted exponents are ordered $i_1 < i_2 < \ldots$, and the notation $[i_j]$ indicates that i_j is omitted from the sequence. The second integrand is obtained from the first by replacing each $f^{i_1,\ldots,i_k}$ by $(\partial f^{i_1,\ldots,i_k}/\partial x^\alpha)dx^\alpha$ and using the manipulation rules for differential forms, including zeroing a product of differentials if a differential is repeated and changing signs if two adjacent differentials are interchanged. This relation between differential forms is indicated by d. Thus, if ω is a k-dimensional form, then

$$\iiint_{\mathfrak{B}} \omega = \iiiint_{\mathfrak{A}} d\omega.$$

This is called the "generalized Stokes' theorem."

The theorems of Gauss and Stokes are used to transform relations between integral expressions into partial differential equations and conversely. Since experimental results must involve quantities obtained by volume or surface integration, this associates the experimental relations with partial differential equations. Techniques for solving partial differential equations provide, therefore, uniform procedures for solving a wide class of problems. Notice that integration can yield relations that are good to a high degree of approximation even when a substance is assumed to have an atomic or molecular fine structure.

Let us discuss some examples of this process. In a field of force F, the work done by the field on a particle that has been moved along the curve $\mathfrak{C}$ is $\int_{\mathfrak{C}}(F^1 dx^1 + F^2 dx^2 + F^3 dx^3)$. If $\mathfrak{C}$ extends between the points P_0 and P_1, then it may be that for any other curve $\mathfrak{C}'$ within a limited distance of $\mathfrak{C}$, the work for the curve $\mathfrak{C}'$ is the same as that for $\mathfrak{C}$. If we take a surface $\mathfrak{S}$ such that $\mathfrak{C}$ lies on $\mathfrak{S}$ and take another curve $\mathfrak{C}'$ on $\mathfrak{S}$ near $\mathfrak{C}$ and also extending from P_0 to P_1, then the combination curve $\mathfrak{C}^*$ consisting of $\mathfrak{C}$ and $\mathfrak{C}'$ in the reverse direction bounds a part of $\mathfrak{S}$, $\mathfrak{B}$. The work done in going around $\mathfrak{C}$ is zero, and thus,

$$0 = \int_{\mathfrak{C}^*} F^\alpha dx^\alpha = \iint_{\mathfrak{B}} (\text{curl } F) \cdot n \, dA.$$

Since $\mathfrak{S}$ is relatively arbitrary, curl F is zero and similarly curl $F=0$ implies that the field work is unchanged by a limited change in the path of integration. This result holds even when the field has a line or lines of singularities such that encircling such a line involves a nonzero amount of work, but the work is unaffected by a variation of $\mathfrak{C}$, which does not cross a line of singularities.

If there are no such lines of singularity, the experimental result that the work is independent of the path is expressed by stating that there is a function ϕ of P_1, (x^1, x^2, x^3), such that

$$\phi(x^1\,x^2, x^3) = \int_{\mathfrak{C}} F^\alpha\,dx^\alpha,$$

and this formula yields grad $\phi = F$. (Changing P_0 changes ϕ only by adding a constant. If there are lines of singularity, ϕ is "multivalued," i.e., assumes one of a number of discrete values.) The necessary and sufficient condition that there be a ϕ such that grad $\phi = F$ is that curl $F = 0$. In terms of differential forms, ϕ is considered to be a differential form of 0 order, that is, of zero degree in the differentials dx^i, and hence grad $\phi = F$ can also be written $d\phi = F^\alpha\,dx^\alpha$. Again where there are no singularities we can differentiate F^α again and we have

$$0 = d(d\phi) = (\text{curl } F)\cdot n\,dA = [\text{curl }(\text{grad }\phi)]\cdot n\,dA$$

$$= \left(\frac{\partial^2\phi}{\partial x^{12}} + \frac{\partial^2\phi}{\partial x^{22}} + \frac{\partial^2\phi}{\partial x^{32}}\right)\cdot n\,dA,$$

and ϕ satisfies a differential equation $\nabla^2\phi = 0$, which can be very useful.

If we are dealing with gravitation the potential due to a single particle is readily computed. This computation is used to compute the potential due to a body of finite density at a point (x^1, x^2, x^3) outside the body. By integrating over the volume containing the body we obtain

$$\phi(x^1, x^2, x^3) = G \iiint \rho(y^1, y^2, y^3)$$

$$\times [(x^1 - y^1)^2 + (x^2 - y^2)^2 + (x^3 - y^3)^2]^{-1/2}\,dy^1\,dy^2\,dy^3$$

$$= G \iiint (\rho/r)dy^1\,dy^2\,dy^3.$$

The force due to the first body on another such body is given by integrating over the second body;

$$F = \iiint \rho(x^1, x^2, x^3)(\nabla_x\phi)dx^1\,dx^2\,dx^3.$$

In the case in which one has a number, n, of rigid spherical bodies in each of which the density is a function of the distance from the center, then the problem of mutual gravitational attraction is the same as that for n particles. This means that one can readily eliminate the expressions for the potentials or forces and obtain a system of simultaneous differential equations on the position coordinates. In the more general case, the problem is resolved into determining the potential function from the distribution of matter. The forces associated with the potential function then determine the movement of the material. We have ignored the finite time of propagation of the gravitational field.

The problems of electrostatics appear at first glance to be similar to those of gravity, but in fact the situation is more complicated. One must deal with the possibility of surface charge as well as a volumetric charge. In addition, one may have "polarization." An uncharged body is the result of a very large number of negative charges canceling the same number of positive charges. A "charge" is usually associated with a relatively small difference between these numbers. However, a small spatial displacement of a much larger number of charges can also produce a field. This is referred to as polarization and occurs in a substance subject to an electric intensity field. The magnetic field has a similar but even more complicated notion of "magnetization." One aspect of these complications is that the expression for the potential may not be unique.

The gradual accumulation of information that finally led to the unified theory of electricity and magnetism occurred over two centuries and involved experimental investigations by Coulomb, Oersted, Ohm, Ampere, Gauss, and Faraday. Our interest is in the use of geometric integration to express the experimental results and in the use of Stokes' theorem to obtain partial differential equations. We present a simplified development to indicate these aspects.

One deals with charges and currents in massive bodies at rest. There is associated with the electric charges a field of force E, the electric intensity, which yields the force on a unit test charge. However, E has been modified by electrostatic polarization from another field, $D = \varepsilon E$, called the electric flux density. It is reasonable to consider ε a constant. Similarly there is a magnetic intensity field H and a magnetic flux field B subject to a matrix relation $B = \mu H$. It is a definite simplification to assume that μ is just matrix multiplication by a constant.

The "flux of force" concept associates with a field of force a "flow" along "lines of force" so that the flux across an element of area $d\sigma$ with area of dA is $F \cdot n\, dA$. If the field F represents an inverse square law, say, of repulsion due to a charge distribution q, then if $\mathfrak{B}$ is the boundary of the region $\mathfrak{A}$, one has the original Gauss result,

$$\frac{1}{4\pi} \iint_{\mathfrak{B}} F \cdot n \, dA = \iiint_{\mathfrak{A}} q \, dV.$$

This yields $(4\pi)^{-1} \operatorname{div} F = q$. One has that D and the charge density ρ are related by $(4\pi)^{-1} \operatorname{div} D = \rho$. This yields for the current J

$$(4\pi)^{-1} \left(\operatorname{div} \frac{\partial D}{dt} \right) = \frac{\partial \rho}{\partial t} = -\operatorname{div} J$$

or

$$\operatorname{div} \left[J + (4\pi)^{-1} \frac{\partial D}{\partial t} \right] = 0.$$

Ampere investigated the magnetic field associated with a current. In particular one can express the work done in moving a unit magnetic pole around a closed curve $\mathfrak{C}$ in terms of J, the current flow through the bounded surface $\mathfrak{B}$:

$$\int_{\mathfrak{C}} H \cdot dx = \iint_{\mathfrak{B}} J \cdot n \, dA.$$

Stokes' theorem then yields curl $H = J$, but since the div (curl H) = 0, this cannot be correct. Maxwell added the term $(1/4\pi)(\partial D/\partial t)$ and the correct equation is

$$\operatorname{curl} H = J + \frac{1}{4\pi} \frac{\partial D}{\partial t}.$$

Faraday investigated the effect of a changing magnetic field on a circuit and showed that a voltage was induced in it according to the relation

$$\int_{\mathfrak{C}} E \cdot dx = -\frac{\partial}{\partial t} \iint_{\mathfrak{B}} B \cdot n \, dA.$$

If the circuit is at rest this yields

$$\operatorname{curl} E = -\frac{\partial B}{\partial t}.$$

We can use the assumptions $\mu H = B$, $D = \varepsilon E$ for μ and ε constants to express the earlier relation as

$$\operatorname{curl} B = \mu J + \frac{1}{c^2} \frac{\partial E}{\partial t},$$

where $c^2 = 4\pi/\mu\varepsilon$.

If one uses curl $E = \text{grad (div } E) - \nabla^2 E$, one can eliminate B to obtain

$$\nabla^2 E - \frac{1}{c^2} \frac{\partial^2 E}{\partial t^2} = \mu \frac{\partial J}{\partial t} + \left(\frac{4\pi}{\varepsilon} \right) \text{grad } \rho.$$

One also has a relationship div $B = 0$ and a similar argument to the above yields

$$\nabla^2 B - \frac{1}{c^2} \frac{\partial^2 B}{\partial t^2} = -\mu \text{ curl } J.$$

One can consider J and ρ as given as in radiation problems, or J and ρ may be determined in terms of E and B by the substance. For example, inside a conductor one would have $J = \sigma E$ and $\rho = 0$.

In order to handle the differential equations, the variations in the force vectors across boundaries between different substances must be known. One has that the tangential components of E and H are continuous across such a boundary and the normal components of D and B are continuous.

The people who developed the theory of electricity and magnetism apparently were motivated by a purely intellectual drive to understand. There is a gap in time between the development of the theory and its application. However, the application has had an enormous cultural effect.

For further discussion of the material in this section, see Abraham and Becker,[1] Bergmann,[2] Mason and Weaver,[16] and Whittaker.[24]

6.7. The Calculus of Variations

In addition to the geometric constructions of Gauss' and Stokes' theorems the calculus of variations also yields problems in differential equations. This area of mathematics arose from the solution of the brachisto-chrone problem by the Bernoullis in the eighteenth century.

Suppose we have an object at a point 0 in a vertical plane. This object can be considered to be a bead that will slide without friction on a wire in the plane to a lower point P_0. The problem is to find the curve for the wire such that the time of descent will be a minimum.

One takes a coordinate system with origin at 0, x axis horizontal, y positive downward, and with P_0 having coordinates (x_0, y_0). Let (x, y) be any point on the arc. The kinetic energy of the object will equal the decline in potential energy,

$$\frac{1}{2} m \left[\left(\frac{dx}{dt} \right)^2 + \left(\frac{dy}{dt} \right)^2 \right] = mgy.$$

If y' is the derivative of y with respect to x, this yields

$$(1+y'^2)\left(\frac{dx}{dt}\right)^2 = mgy.$$

Thus,

$$\frac{1}{(2g)^{1/2}}\left(\frac{1+y'^2}{y}\right)^{1/2} dx = dt,$$

and the time of descent is

$$T = \frac{1}{(2g)^{1/2}}\int_0^{x_0}\left(\frac{1+y'^2}{y}\right)^{1/2} dx.$$

This is a problem of the following type. One has a function of three variables $F(x, y, y')$. We must choose a function y with given values at 0 and x_0 that will minimize the integral

$$I = \int_0^{x_0} F(x, y, y')dx.$$

Suppose y is a solution to this problem and δy is any function with a continuous derivative that is zero at 0 and x_0. Let

$$I(\varepsilon) = \int_0^{x_0} F(x, y+\varepsilon\,\delta y, y'+\varepsilon\,\delta y')dx.$$

Then $I(\varepsilon)$ must have a minimum for $\varepsilon = 0$ and $dI/d\varepsilon = 0$ at $\varepsilon = 0$. Formally we have

$$0 = \int_0^{x_0}\left(\frac{\partial F}{\partial y}\,\delta y + \frac{\partial F}{\partial y'}\,\delta y'\right)dx$$

$$= \int_0^{x_0}\frac{\partial F}{\partial y}\,\delta y\,dx + \frac{\partial F}{\partial y'}\,\delta y\bigg|_0^{x_0} - \int_0^{x_0}\frac{d}{dx}\left(\frac{\partial F}{\partial y'}\right)\delta y\,dx$$

$$= \int_0^{x_0}\left[\frac{\partial F}{\partial y} - \frac{d}{dx}\left(\frac{\partial F}{\partial y'}\right)\right]\delta y\,dx.$$

This last equation is satisfied by all differentiable δy that satisfy the end conditions. If y is such that $\partial F/\partial y - (d/dx)(\partial F/\partial y')$ is continuous as a function of x, this is only possible if

$$\frac{\partial F}{\partial y} - \frac{d}{dx}\left(\frac{\partial F}{\partial y'}\right) = 0.$$

This is called the Euler equation for the problem. If $\partial^2 F/\partial y'^2$ is not zero, this is a second-order differential equation, which is a necessary condition for the solution to the problem. The theory investigates sufficient conditions.

In our example we can take

$$F = [(1 + y'^2)/y]^{1/2}.$$

One can show that the Euler equation for this problem implies

$$\frac{2y''}{1 + y'^2} + \frac{1}{y} = 0.$$

By multiplying by y', this equation can be integrated to yield

$$(1 + y'^2)y = 2a.$$

This first-order differential equation can be integrated and the solution, which at $x = 0$ has the value $y = 0$, has the parametric form

$$y = a(1 - \cos \theta), \qquad x = a(\theta - \sin \theta).$$

One can also show that $\theta = (g/a)^{1/2}t$. This curve is also called the "cycloid." The quantity a must be determined so that the "brachistochrone" goes through (x_0, y_0).

The calculus of variations was applicable to a considerable number of interesting geometric and physical problems. It is still important in many practical situations such as optimal control. It structured the Lagrangian and Hamiltonian formulations of dynamics.

From the mathematical point of view its importance lay in the fact that it dealt with sets of functions rather than, say, sets of numbers. We have minimum problems in which one chooses the best function rather than the most appropriate value of a variable. This led to Volterra's theory of functions of curves, which was the predecessor of the modern theory of function spaces and abstract topological spaces.

For further discussion of the material in this section, see Bliss.[3]

6.8. Dynamics

Newton's dynamics was formulated in terms of a Galilean system of coordinates. The spatial coordinates are therefore Euclidean, which is quite restrictive, and we will discuss the formulation of mechanics due to Lagrange, which permits a much more general choice of spatial coordinates.

We begin with a Newtonian formulation. In order to deal with an extensive body, one must consider it to be a collection of particles. We have for the Euclidean coordinates a system of differential equations $m_j x_j^i = F_j^i$, $i = 1, 2, 3$, $j = 1, \ldots, N$. It is convenient to consider a single superscript with range $i = 1, \ldots, 3N$. In the case of a rigid body there are many relations between the coordinates of the particles that always hold. The forces are such that these relations appear as integrals of these equations. If we have R such relations between the coordinates, we can make a change of variables so that R of the new coordinates are constants corresponding to these R restraints, and there are n new variables $q^1, ..., q^n$ with $n = 3N - R$. However, there may be other relations between the new variables that do not correspond to a functional relation between coordinates, for example, a relation between the differentials of these variables such as $A_{i\alpha}dq^\alpha = 0$ or a kinetic relation such as $A_{i\alpha}\dot{q}^\alpha = 0$ (see Whittaker[23]). A relation between differentials, $A_{i\alpha}dq^\alpha = 0$, corresponds to a functional relation $F(q^1, ..., q^n) = c$ only if the A_{ij} values are proportional to $\partial F/\partial x^j$. A relation that is not equivalent to a functional relation is called nonholonomic, and a mechanical system that is not subject to any nonholonomic relation is termed holonomic.

The forces F_i at a point can be represented relative to the differential vector space of the (dx^i) as corresponding to a linear functional with value $F_\alpha dx^\alpha$. If we perform a theoretical displacement indicated by the operator δ, $W(\delta) = F_\alpha \delta x^\alpha$ is the "virtual work" corresponding to the "virtual displacement."

The Newtonian equations $m_i\ddot{x}^i = F_i$ can be transformed into a form that is suitable for a change of variable. One introduces the kinetic energy, $T = \frac{1}{2}m_\alpha \dot{x}^{\alpha 2}$, for which

$$m_i\ddot{x}^i = \frac{d}{dt}\left(\frac{\partial T}{\partial \dot{x}^i}\right) - \frac{\partial T}{\partial x^i},$$

where the partial derivatives correspond to T as a function of x^i and $\dot{x}^i$ and consequently $\partial T/\partial \dot{x}^i = m_i\dot{x}^i$ and $\partial T/\partial x = 0$. A history of the mechanical system corresponds to a curve, $\mathfrak{C}$, $x^i = x^i(t)$ in the $3N$-dimensional x space. The original equation $F_i - m_i\ddot{x}^i = 0$ can now be written

$$F_i + \frac{\partial T}{\partial x^i} - \frac{d}{dt}\left(\frac{\partial T}{\partial \dot{x}^i}\right) = 0.$$

We introduce a "virtual displacement" of $\mathfrak{C}$ in the manner of the calculus of variations, i.e., with zero displacement at the terminal points. If we multiply the above equation by δx^i and sum over i and integrate along the curve $\mathfrak{C}$, we get two integral terms whose sum is zero. One of these, $\int F_\alpha \delta x^\alpha dt = \int W(\delta) dt$, has an invariant integrand under changes of the space variables. (The parameter t is also the time and plays a special role.) The other term is

$$\int_{\mathfrak{C}}\left[\frac{\partial T}{\partial x^\alpha} - \frac{d}{dt}\left(\frac{\partial T}{\partial \dot{x}^\alpha}\right)\right]\delta x^\alpha dt = \int_{\mathfrak{C}}\left(\frac{\partial T}{\partial x^\alpha}\delta x^\alpha + \frac{\partial T}{\partial \dot{x}^\alpha}\delta \dot{x}^\alpha\right)dt = \delta\int_{\mathfrak{C}} T\, dt$$

by the usual argument of the calculus of variations, in which one integrates the second term by parts using $(d/dt)\delta x^\alpha = \delta(d/dt)x^\alpha = \delta\dot{x}^\alpha$ and the fact that $\delta x^\alpha = 0$ at the terminal points of $\mathfrak{C}$. Thus, our original Newtonian equations

$m_i \ddot{x}^i = F^i$ imply

$$\int W(\delta)\, dt + \delta \int T\, dt = 0,$$

which is invariant under changes in the spatial or kinetic variables, and the calculus of variations will yield the converse result.

We can use the restraints that can be expressed in terms of the coordinates and time to introduce a new system of spatial coordinates, R of which are constants for the motion. The remaining variables $q^1, \ldots, q^n$ now determine an n-dimensional subspace of the original $3N$-dimensional manifold by equations

$$x^i = x^i(q^1, \ldots, q^n, t).$$

(The time variable t is the same before and after the change of coordinates.) The curve $\mathfrak{C}$ is a path in this subspace so that there exists a preimage $\mathfrak{C}^*$ in the q space. The virtual displacements that satisfy the restraint restrictions are also in this subspace and hence correspond to virtual displacements of $\mathfrak{C}^*$ in the q space. In our original variation result, we can make the change of variables from the $3N$ x^i and suppress the new variables, which are constant because of the restraints. The result is a variation statement on $\mathfrak{C}^*$,

$$\int_{\mathfrak{C}} W(\delta)\, dt + \delta \int_{\mathfrak{C}} T\, dt = 0.$$

For further discussion of the material in this section, see Whittaker.[23]

If there are no nonholonomic restraints, this is now just a calculus of variations problem on the $q^1, \ldots, q^n$. Since it is the q^i as functions of the time that are usually desired, this represents an elimination of the awkward restraint relations. If one has nonholonomic restraints, for example, $A_{i\alpha}\delta x^\alpha = 0$, these restraints are transformed by the change of variables into another form, for example, $B_{i\alpha}\,\delta q^\alpha = 0$, since one can then consider the variation of the restraint variables as zero. The variation problem can then be considered as an extremal problem in the calculus of variations subject to auxiliary restraints. In any case we have a formulation of mechanics of great technical value that permits transformations of the q variables.

The value of this procedure lies in the fact that both T and $W(\delta)$ can be expressed in terms of the $q^1, \ldots, q^n$. Thus,

$$T = \frac{1}{2} m_\alpha \left(\frac{dx^\alpha}{dt}\right)^2 = \frac{1}{2} m_\alpha \left(\frac{\partial x^\alpha}{\partial q^\beta}\, \dot{q}^\beta + \frac{\partial x^\alpha}{\partial t}\right)^2$$

and

$$W(\delta) = F_\alpha\, \delta x^\alpha = F_\alpha \frac{\partial x^\alpha}{\partial q^\beta}\, \delta q^\beta = (\text{say})\ Q_\beta\, \delta q^\beta.$$

(The time is not varied in the virtual displacements.) In practice one would try to express T and $W(\delta)$ directly in terms of the q^i, as for example in the case of a pendulum. There is a special case in which $W(\delta)$ can be expressed as the variation of a potential function, i.e., $W(\delta) = -\delta W = (\partial W/\partial q^\alpha)\,\delta q^\alpha$, in which case the variation problem can be expressed as

$$\delta \int (-W+T)dt = 0.$$

Another special case is that in which the "generalized forces," the Q_i, are linear in the q^i; for example, $Q_i = Q_{i,0} + a_{i\alpha}q^\alpha$. This situation can be interpreted in the case where the a_i are constants. If the matrix (a_{ij}) is symmetric and negative definite, one is dealing with a dissipation of energy, as, for example, the dissipation due to resistance in an electric circuit. On the other hand a constant magnetic field acting on a moving charged particle will yield a situation in which the a_{ij} are *antisymmetric*.

When one has a potential function, W, we have the Lagrangian $L = T - W$ and the Lagrangian equations from the variational principle

$$\frac{d}{dt}\left(\frac{\partial L}{\partial \dot{q}^i}\right) - \frac{\partial L}{\partial q^i} = 0.$$

This is a system of n second-order differential equations on the n functions $q^1(t), \ldots, q^n(t)$. There is considerable theoretical interest in an alternate formulation in terms of the momenta, $p_i = \partial L/\partial \dot{q}^i$, regarded as functions of the time. These constitute n linear equations on the $\dot{q}^i$ and thus can be solved to express the $\dot{q}_i$ as linear combinations of the p_i. Consider

$$H(p, q) = p_\alpha \dot{q}^\alpha - L(\dot{q}, q).$$

Here the $\dot{q}^i$ on the righthand side are considered to be functions of p_i and q_j. Consequently

$$\frac{\partial H}{\partial q^i} = p_\alpha \frac{\partial \dot{q}^\alpha}{\partial q^i} - \frac{\partial L}{\partial q^i} - \frac{\partial L}{\partial \dot{q}^\alpha}\frac{\partial \dot{q}^\alpha}{\partial q^i} = -\frac{\partial L}{\partial q^i}$$

$$\frac{\partial H}{\partial p_i} = \dot{q}^i + p^\alpha \frac{\partial \dot{q}^\alpha}{\partial p_i} - \frac{\partial L}{\partial \dot{q}^\alpha}\frac{\partial \dot{q}^\alpha}{\partial p_i} = \dot{q}^i.$$

Thus, the original system of differential equations is equivalent to the $2n$ equations

$$\dot{p}_i = -\frac{\partial H}{\partial q^i}, \qquad \dot{q}^i = \frac{\partial H}{\partial p^i}$$

in which the right-hand side is expressed in terms of the p_i and q_i and the system is explicitly solved for the derivatives.

The exploration of the use of the Hamiltonian is of course part of the theory of classical dynamics. This includes the role of the Hamilton–Jacobi equation, which is a first-order partial differential equation in the form $H(\partial S/\partial q^i, q^i) = 0$ on an unknown function $S(q^1, ..., q^n)$. One important aspect is the study of canonical transformations, i.e., those transformations on the $2n$-dimensional p, q space that still yield equivalent dynamic problems.

6.9. Manifolds

It is natural to describe the behavior of substances in terms of a rather general three-dimensional geometry. However, the evolution of the theory of dynamics in the hands of Euler, Lagrange, Hamilton, and Jacobi proceeded in a much more general n-dimensional framework, paralleling the development of appropriate mathematical tools such as the theory of differential equations and the calculus of variations. In the nineteenth century the analytic formulation of geometric ideas crystallized into the concept of invariant structures associated with permissible transformation groups, leading to Riemannian and more general geometries. These ideas are particularly appropriate for dynamics and eventually provided remarkable flexibility in the formulation of modern theories such as relativity.

Let us consider in an informal way the modern notion of a differentiable manifold. A differential manifold of n dimensions is a point set or "space" that is smooth enough so that at each point one can define a "differential vector space." The problem is to set this up in an analytic manner. There are many formal procedures for doing so that are logically satisfactory and provide little insight. Let us describe the setup informally. Consider the manifold as being divided into a number of overlapping patches. Each patch is in a one-to-one correspondence with an open unit cube in n-dimensional space. Since each point P_0 is in at least one patch, each P_0 has at least one set of coordinates $x^1, ..., x^n$. If P_0 is in the overlap of two patches and has two sets of coordinates $x^1, ..., x^n$ and $y^1, ..., y^n$, then there is an open neighborhood of $x^1, ..., x^n$ that is in a one-to-one correspondence with an open neighborhood of $y^1, ..., y^n$ given by n functions, $y^i = f^i(x^1, ..., x^n)$, which are indefinitely differentiable. Thus, if the assignment of coordinates yields more than one set, these different assignments have a local relationship that can be differentiated any number of times. We refer to the mapping of a patch of the manifold onto the unit n cube as a "chart" and the whole procedure including the overlap functions as an "atlas."

We suppose one such atlas as given. Now consider another such atlas such that if a point P on the manifold has coordinates $x^1, ..., x^n$ from a chart of the first atlas and coordinates $y^1, ..., y^n$ from a chart of the second atlas, then again we can find an open neighborhood of $x^1, \ldots, x^n$ that is mapped into an open neighborhood of $y^1, \ldots, y^n$ by this relation in terms of a set of n indefinitely differentiable functions, and the inverse mapping has the same property. This relationship between two such atlases is symmetric and transitive and one can consider the collection of atlases that are equivalent to the given one by this relation.

Such a collection of atlases is a global structure on our manifold that can be used to specify a "geometry." For it is clear that any one-to-one mapping of the basic point set of the manifold onto itself will take an atlas into an atlas. The group of transformations that determines the geometry is the set of such mappings that preserves the collection. It is intuitively clear that the local geometric properties of the manifold are those that are preserved by passing from one chart to another, i.e., those preserved by the f^i mappings given above.

These charting constructions imply the existence of a "differential vector space" at each point P_0 of the manifold. In other words, the geometry of manifolds includes such a space, since this space can be shown to be invariant under the group transformations. The invariance is for the space as an entity, not for the individual vectors. To show this we consider a function $F(P)$ of the points of the manifold. For any chart that contains P_0, there is a neighborhood of the coordinates $x_0^1, \ldots, x_0^n$ such that for points P with coordinates in this neighborhood $x^1, \ldots, x^n$, $F(P) = F(x^1, \ldots, x^n)$. Suppose F is indefinitely differentiable. Then at P_0, $dF = (\partial F / \partial x^\alpha) \, dx^\alpha$ for any differential vector $(dx^1, \ldots, dx^n)$. If we make the change of variable $y^i = f^i(x^1, \ldots, x^n)$, the differential vector is transformed by the equation $dy^i = a_\alpha^i \, dx^\alpha$, where $a^i = \partial y^i / \partial x^\alpha$. Thus, the space of such differential vectors is an affine space.

The analytic expressions for geometric notions in this affine space involve numerical arrays called "tensors." When the reference coordinates are changed as in the preceding paragraph, these arrays are transformed by means of the $a_j^i = \partial y^i / \partial x^j$ and the $A_j^i = \partial x^i / \partial y^j$. Thus, the vector $(dx^1, \ldots, dx^n)$ becomes $(dy^1, \ldots, dy^n)$, where $dy^i = a_\alpha^i \, dx^\alpha$. This latter relation also implies $dx^i = A_\alpha^i \, dy$, which will specify the transformation of the coefficients of differential forms. Thus, the differential form $c_{\alpha\beta} \, dx^\alpha \, dx^\beta$ becomes $c'_{\alpha\beta} \, dy^\alpha \, dy^\beta$ for $c'_{ij} = c_{\alpha\beta} A_i^\alpha A_j^\beta$. The components of tensors are usually expressed with subscripts and superscripts. For example, the component of a tensor t may be expressed as t_{jk}^i, which indicates that the transformation rule is $t_{jk}^{i\,\prime} = a_\alpha^i t_{\beta\gamma}^\alpha A_j^\beta A_k^\gamma$.

The relations expressible in terms of this infinitesmal affine geometry are invariant under the larger set of coordinate transformations. One can formulate the problem of establishing global consequences of local relations. Less generally, integration concepts can be introduced in the form of integrals over k-dimensional submanifolds of our original n-dimensional manifold. These integrals have integrands which are kth order exterior forms over the n-dimensional differential vector space. The k-dimensional submanifold must be divided into pieces each of which corresponds to a k-dimensional subset of the n cube of a specific chart. For each such piece the numerical interpretation of the integral is clear. However, the integrand must have a certain tensorial significance if the result is to be independent of the choice of the chart. Thus, the integral

$$\iiint P_{\alpha\beta\gamma}\, dx^\alpha\, dx^\beta\, dx^\gamma$$

is a scalar invariant if the P_{ijk} are the components of a covariant tensor.

For further discussion of the material in this section, see Helgason[11] and Whitney.[22]

6.10. The Weyl Connection

Tensor invariance corresponds to the least specific structures on the manifold. An additional type of structure is based on a construction due to H. Weyl that compares the differential vector space at a point P with that at any "infinitesimally displaced" point Q. The vector space E_p is related to the equivalent space E_Q by a linear transformation G, which depends on $\overrightarrow{PQ}$. Thus, if $(dx^1,...,dx^n)$ is a differential vector at P, then $G(dx^1,...,dx^n)$ is a differential vector at Q, $(x^1+\delta x^1, ..., x^n+\delta x^n)$. It is reasonable to assume that G can be described by a matrix

$$G = \{\delta^i_j + \Gamma^i_{j\alpha}\,\delta x^\alpha\} = I + \Gamma_\alpha\,\delta x^\alpha,$$

where I and Γ_α are transformations given by the appropriate matrices.

One must also specify the situation relative to overlap. Suppose at P we have another chart and coordinate system that assigns to P the coordinates $y^1, ..., y^n$ and Q the coordinates $y^1+\delta y^1, ..., y^n+\delta y^n$. Then $y^i = f^i(x^1, ..., x^n)$ and $dy^i = a^i_\alpha\, dx^\alpha$ for $a^i_\alpha = \partial f^i/\partial x^\alpha$. We use the obvious notation $dy = a\, dx$, $dx = A\, dy$ for $A^i_j = \partial x^i/\partial y^j$. Let the bar indicate a reference to the point Q. Then $d\bar{y} = \bar{a}\, d\bar{x} = (a+\delta a)d\bar{x} = (a+\Delta_\alpha\,\delta x^\alpha)d\bar{x}$, where Δ_k is the matrix $\partial^2 y^i/\partial x^j\,\partial x^k$ where $i,j = 1, ..., n$. Thus,

$$d\bar{y} = (a+\Delta_\alpha\,\delta x^\alpha)d\bar{x} = (a+\Delta_\alpha\,\delta x^\alpha)G\, dx = (a+\Delta_\alpha\,\delta x^\alpha)(I+\Gamma_\beta\,\delta x^\beta)dx$$

$$= (a+\Delta_\alpha\,\delta x^\alpha)(I+\Gamma_\beta\,\delta x^\beta)A\, dy = dy + (\Delta_\alpha\,\delta x^\alpha)A\,\delta y + a(\Gamma_\beta\,\delta x^\beta)A\, dy + \cdots.$$

If we use tensor rather than matrix notation this becomes

$$d\bar{y}^i = dy^i + \frac{\partial^2 y^i}{\partial x^\rho x^\sigma} A_\alpha^\rho A_\beta^\sigma \, \delta y^\alpha \, dy^\beta + a_\tau^i \Gamma_{\rho\sigma}^\tau A_\alpha^\rho A_\beta^\sigma \, \delta y^\alpha \, dy^\beta$$

$$= dy^i + \left(\frac{\partial^2 y^i}{\partial x^\rho \, \partial x^\sigma} + a^i \Gamma_{\rho\sigma}^\tau \right) A_\alpha^\rho A_\beta^\sigma \, \delta y^\alpha \, dy^\beta,$$

and if the prime indicates the connection for the y^i any $\bar{y}^i$,

$$\Gamma_{j,k}^{\prime i} = \left(\frac{\partial^2 y^i}{\partial x^\rho \, \partial x^\sigma} + a_\tau^i \Gamma_{\rho\sigma}^\tau \right) A_j^\rho A_k^\sigma.$$

Of course, the change of charts relative to a change of atlas yields the same relation. One can show that if one makes a further change of coordinates from y to z, one gets the same relation between the Γ_{jk}^i for the z^i and $\bar{z}^i$ and the original Γ_{jk}^i for the x^i and $\bar{x}^i$ as one would obtain if one made a direct change from x to z, eliminating the intermediate transformation to the y coordinates. Thus, if the Γ_{jk}^i are given in one coordinate system, they are uniquely determined for any other coordinate system and have a unique transformation relationship relative to change of charts. The Γ_{jk}^i can therefore be introduced as a further geometric structure beyond the tensor relations.

Because of the second-order partial derivative term in the transformation formula, the Γ_{jk}^i are not components of a tensor. However, for a change of coordinates that is given locally by a linear matrix with constant elements, the transformation rule is a tensor rule. In particular if one can find a coordinate system for which the Γ_{jk}^i are zero, then the space has a local affine character. In general, if we are given a coordinate system $x^1, \ldots, x^n$ with a corresponding set of Γ_{jk}^i, we can find another coordinate system $y^1, \ldots, y^n$ with $\Gamma_{jk}^{\prime i} = 0$ if we can solve the system of equations

$$\frac{\partial y^i}{\partial x^\alpha} \Gamma_{jk}^\alpha + \frac{\partial^2 y^i}{\partial x^j \, \partial x^k} = 0.$$

These equations imply

$$\Gamma_{jk}^i = \Gamma_{kj}^i \qquad \text{and} \qquad \frac{\partial \Gamma_{jk}^i}{\partial x^l} - \frac{\partial \Gamma_{jl}^i}{\partial x^k} + \Gamma_{\alpha l}^i \Gamma_{jk}^\alpha - \Gamma_{\alpha k}^i \Gamma_{jl}^\alpha = 0.$$

The relation between the differential vector space of "infinitesimally close" points implies a relation between the differential vector spaces of points that can be joined by an appropriate curve $\mathfrak{C}$. Such a curve would consist of a finite sequence of adjoined arcs $\mathfrak{C}_i$, each of which is in the patch of the manifold associated with a chart. The relation in general, then, is implied by the relations between the vector spaces of the endpoints of such a chart.

For the coordinate system associated with the chart a specific arc can be represented by the parametric equations $x^i = X^i(t)$ for $0 \leqslant t \leqslant 1$. To avoid confusion let us represent differential vectors by single letters, i.e., by $(\delta^1,...,\delta^n)=\delta$ and if one such vector is chosen at every point on the curve $P(t)$, we can denote it $\delta(t)=(\delta^1(t),\delta^2(t),...,\delta^n(t))$. Suppose we consider $\delta(t)$ as being moved along the curve in accordance with the connection, i.e.,

$$\delta^i(t+dt)=\delta^i(t)+\Gamma^i_{\alpha\beta}\,\delta^\alpha\,\frac{dX^\beta}{dt}\,dt$$

or

$$\frac{d}{dt}(\delta^i)=\Gamma^i_{\alpha\beta}\,\delta^\alpha\,\frac{dX^\beta}{dt}\,.$$

This is a system of differential equations and a vector associated with the solution $(\delta^1(t),\delta^2(t),...,\delta^n(t))$ can be considered to be the transported vector of the corresponding initial conditions $(\delta^1_0,...,\delta^n_0)$ at the initial point of the arc.

For further discussion of the material in this section, see Brillouin,[5] Helgason,[11] and Whitney.[22]

6.11. The Riemannian Metric

The invariance of tensor relations corresponds to the affine character of the differential vector space, and the connection Γ^i_{jk} represents additional geometric structure. The affine character of a linear space can be specialized into a Euclidean space by designating a tensor g_{ij} for each point on the manifold. Such a tensor, g_{ij}, determines a connection Γ^i_{jk} by the condition that the transport operation preserves the inner product relation. Let η and ξ denote two differential vectors at a point $x^1,...,x^n$ and consider the transport of $g_{\alpha\beta}\eta^\alpha\xi^\beta$ to the point $x^1+\delta^1,...,\delta x^n+\delta^n$. Then

$$0=\delta(g_{\alpha\beta}\eta^\alpha\xi^\beta)=(\delta g_{\alpha\beta})\eta^\alpha\xi^\beta+g_{\alpha\beta}(\delta\eta^\alpha)\xi^\beta+g_{\alpha\beta}\eta^\alpha\delta\xi^\beta$$

$$=\frac{\partial g_{\alpha\beta}}{\partial x^\gamma}\delta x^\gamma\eta^\alpha\xi^\beta+g_{\alpha\beta}\Gamma^\alpha_{\rho\sigma}\,\delta x^\rho\,\eta^\sigma\xi^\beta+g_{\alpha\beta}\Gamma^\beta_{\rho\sigma}\,\delta x^\rho\,\xi^\sigma\eta^\alpha$$

$$=\left(\frac{\partial g_{\alpha\beta}}{\partial x^\gamma}+g_{\sigma\beta}\Gamma^\sigma_{\gamma\alpha}+g_{\alpha\sigma}\Gamma^\sigma_{\gamma\beta}\right)\eta^\alpha\xi^\beta\delta x^\gamma.$$

Thus the connection is determined by the relations

$$\frac{\partial g_{ij}}{\partial x^k}+g_{\alpha j}\Gamma^\alpha_{ki}+g_{i\alpha}\Gamma^\alpha_{kj}=0\,.$$

The g_{ij} and the Γ^i_{jk} are considered to be symmetric in the subscripts. If this equation is written three times with cyclic permutations of the subscript, that is, first for i, j, k, then for j, k, i, and then for k, i, j, and the first equation is subtracted from the sum of the other two, one obtains

$$\left(\frac{\partial g_{jk}}{\partial x^i}+\frac{\partial g_{ki}}{\partial x^j}-\frac{\partial g_{ij}}{\partial x^k}\right)+2g_{\alpha k}\Gamma^\alpha_{ij}=0.$$

Since we assume that the matrix of the g_{ij} is not singular, this determines the connection Γ^i_{jk}. The connection associated with the "metric tensor" g_{ij} is said to correspond to a "parallel displacement" of the differential vector.

Along an arc located within a given chart, the parallel displacement of a vector is given by

$$\frac{d}{dt}(\delta^i)=\Gamma^i_{\alpha\beta}\,\delta^\alpha\,\frac{dX^\beta}{dt}\,.$$

This equation is valid for any choice of the parameter t. Since $g_{\alpha\beta}\,dx^\alpha\,dx^\beta$ is now an invariant for the geometry if the arc is such that

$$\frac{ds}{dt}=\left(g_{\alpha\beta}\frac{dX^\alpha}{dt}\frac{dX^\beta}{dt}\right)^{1/2}$$

is not zero along the curve, we can introduce s as a parameter and consider the curves obtained by displacing the tangent vector in its own direction. This would yield the system of differential equations obtained by replacing the δ_i by dx^i/ds and t by s,

$$\frac{d^2x^i}{ds^2}=\Gamma^i_{\alpha\beta}\frac{dx^\alpha}{ds}\frac{dx^\beta}{ds}\,.$$

The system of curves that satisfy these equations are the "geodesics." This refers to the fact that such a curve is an extremal for the integral

$$\int ds=\int\frac{ds}{dt}\,dt=\int\left(g_{\alpha\beta}\frac{dx^\alpha}{dt}\frac{dx^\beta}{dt}\right)^{1/2}dt=(\text{say})\int T\,dt.$$

The Euler equations for this variation,

$$\frac{d}{dt}\left(\frac{\partial T}{\partial\dot x^i}\right)=\frac{\partial T}{\partial x^i}\,,$$

becomes

$$\frac{d}{dt}\frac{1}{(g_{\alpha\beta}\dot x^\alpha\dot x^\beta)^{1/2}}(g_{i\alpha}\dot x^\alpha)=\frac{1}{(g_{\alpha\beta}\dot x^\alpha\dot x^\beta)^{1/2}}\frac{\partial g_{\alpha\beta}}{\partial x^i}\,\dot x^\alpha\dot x^\beta.$$

Since $dt/ds=(g_{\alpha\beta}\dot x^\alpha\dot x^\beta)^{-1/2}$, we can multiply these equations by dt/ds and obtain

$$\frac{d}{ds}\left(g_{i\alpha}\frac{dx^\alpha}{ds}\right)=\frac{1}{2}\frac{\partial g_{\alpha\beta}}{\partial x^i}\frac{dx^\alpha}{ds}\frac{dx^\beta}{ds}\,.$$

This yields

$$g_{i\alpha} \frac{d^2 x^\alpha}{ds^2} = \frac{1}{2}\left(\frac{\partial g_{\alpha\beta}}{\partial x^i} - \frac{\partial g_{i\alpha}}{\partial x^\beta} - \frac{\partial g_{i\beta}}{\partial x^\alpha}\right)\frac{dx^\alpha}{ds}\frac{dx^\beta}{ds},$$

and this is equivalent to the geodesic equations.

An extremal condition is, of course, independent of the choice of the coordinates. The principles of general relativity requires that the laws of physics be independent of the choice of the coordinates used to express space and time. Thus, in modern theories of motion the space–time continuum is given a Riemannian geometric structure and the motion of an "infinitesimal particle" is described as moving on a geodesic. This means that the second-order Newtonian equations are replaced by the geodesic equations, but the Newtonian equations must be an excellent approximation when the speed of the particle is small relative to the speed of light. Instead of "forces" in the new equations, we have Γ terms with geometric significance. This has yielded a more precise description of gravitation in the solar system.

But there are other forces besides gravitation, and efforts have been made to describe them by using the more general notion of a connection or Γ concept. For if the Γ^i_{jk} are given, there will not be in general a tensor g_{ij} that satisfies the relation

$$\frac{\partial g_{ij}}{\partial x^k} + g_{i\alpha}\Gamma^\alpha_{kj} + g_{j\alpha}\Gamma^\alpha_{ki} = 0,$$

since one can readily obtain necessary conditions on the Γ^i_{jk} that are not always fulfilled. Those developments are discussed in Brillouin,[5] Lanczos, and Weyl.[21]

Thus, the development of this analysis has produced subtle and beautiful intellectual concepts. For further discussion of the material in this section, see Eisenhart.[7]

For further discussion of the material in this chapter, consult Heath,[9] Heitler,[10] Kellogg,[12] Lamb,[14] Van der Waerden,[20] and Whittaker.[23]

Exercises

6.1. By means of summation formulas, show that the area under $y = x^n$ between 0 and a is $a^{n+1}/(n+1)$.

6.2. The "principle of Cavalieri" is trivial in terms of modern theories of integration. But it is interesting to see what is required to validate the argument given by Cavalieri (see Smith,[19] p. 605), e.g., in the case of areas.

6.3. Compare the geometrical arguments of Section Two of Book One of Newton's *Principia*[6] with the corresponding modern analytic arguments.

6.4. How does

$$\tfrac{1}{4} = 1 - 2 + 3 - 4 + \cdots$$

arise by interchanging limits? Evaluate the partial sums of the power series $1 - 2x + 3x^2 - 4x^3 + \cdots$. What happens as x approaches 1? What is the effect of introducing Cesaro summation? In regard to multiplying series term by term, what is the relationship between the square of a partial sum for $1/(1+x)$ and a remainder and a partial sum for $1/(1+x)^2$ and its remainder?

6.5. Explore the geometric theory of the shape of gear teeth.

6.6. For a curve in three dimensions given in parametric form, determine the oscillating plane, the curvature, and the normal.

6.7. Let $v^i(x^1, x^2, x^3, t)$ denote the velocity of the substance at the point (x^1, x^2, x^3) at time t. We take a fixed time t_0 as a reference time. The point in the substance that at t_0 was at x_0^1, x_0^2, x_0^3 is at time t at the point x^1, x^2, x^3. We have then a transformation on the spatial coordinates for each time value t given by $x^i = x^i(x_0^1, x_0^2, x_0^3, t)$, which is the solution of the system of differential equations

$$\frac{dx^i}{dt} = v^i(x^1, x^2, x^3, t),$$

which at $t = t_0$ takes on the value x_0^1, x_0^2, x_0^3. How are the partial derivatives $\partial x^i / \partial x_0^j$ determined? How is an integral

$$\iint (f^1 \, dx^2 \, dx^3 + f^2 \, dx^3 \, dx^1 + f^3 \, dx^1 \, dx^2)$$

expressed in terms of x_0^1, x_0^2, x_0^3?

6.8. State the implicit function theorem for 3 or n variables. What is the condition for "functional dependence"? What is the formula for a change of independent variables in a multiple definite integral?

6.9. What are "existence theorems" for systems of differential equations? What are "uniqueness theorems"? Give examples of each type.

6.10. Show that for an incompressible substance $D\rho/Dt = 0$.

6.11. At one time, heat Q was considered a fluid and the temperature T in a homogeneous substance was assumed to be proportional to the local density of heat fluid. Thus, Q was an extensive quantity related to the intensive quantity T by

$$Q = \iiint_{\mathfrak{A}} CT \, dV,$$

where the constant C is the "specific heat." The rate of flow of Q was considered to be proportional to the negative temperature gradient. If $\mathfrak{B}$ is the surface bounding the region $\mathfrak{A}$ and n is as usual the outward normal, then this assumption can be written

$$\frac{dQ}{dt} = \iint_{\mathfrak{B}} k(\text{grad } T) \cdot n \, dA,$$

where k is the "heat conductivity." Since this equation is valid for an arbitrary region $\mathfrak{A}$, these assumptions lead to the "equation of heat":

$$\frac{\partial^2 T}{\partial x^{12}} + \frac{\partial^2 T}{\partial x^{22}} + \frac{\partial^2 T}{\partial x^{32}} = a^2 \frac{\partial T}{\partial t}$$

with $a^2 = k/C$.

6.12. A k-dimensional subspace in n-dimensional space is given by n equations $x^i = x^i(r^1, \ldots, r^k)$. If in the differential form $\Sigma f^{i_1, \ldots, i_k} dx^{i_1} \ldots dx^{i_k}$ we substitute $dx^i =$

$(\partial x^i/\partial r^\alpha)dr^\alpha$ and use the manipulation rules, what is the integrand for integration relative to $r^1, ..., r^k$?

6.13. Prove Stokes' and Gauss' theorems. How are the rules for differentiating "differential forms" applicable? (See Flanders.[8])

6.14. How is the area of a patch of surface $\iint dA$ expressed in terms of the vectors R and S? How is this related to the Archimedean definition of area of a surface?

6.15. Show that a vector field F can be expressed in the form $G + H$, where curl $G = 0$ and div $H = 0$. Prove that if curl $G = 0$, then $G = \text{grad } \phi$ for a function of x^1, x^2, x^3. Prove that if div $H = 0$, then there is a vector field A such that curl $A = H$. (This is a very standard theorem that is used in practically every theory. It is suggested that if the student is not familiar with it, he try to establish it on his own before looking it up in one of the standard books.)

6.16. Obtain Stokes' theorem and Gauss' theorem from the formula for the generalized Stokes' theorem. Also show that $d(d\omega) = 0$.

6.17. Obtain the potential function of a thin spherical shell and show that for a point outside the shell it is the same as that of a particle of the same mass located at the center. What is the potential function inside the shell and what is the significance of this result? What is the potential function of a rigid sphere with density proportional to the distance to the center. What is the energy of a configuration of two such bodies relative to when they are at an infinite distance? What is the attractive force between them?

6.18. Consider the contribution to the potential ϕ at the point $P(x^1, x^2, x^3)$ of a number of charges, e_i, in a small region, $\mathfrak{A}$ around a point Q. The diameter of $\mathfrak{A}$ is small relative to $r = PQ$. Suppose e_i is at P_i, l_i is the distance P_iQ and $\angle P_iQP = \theta_i$. Then

$$r_i = \text{dist}(P_iP) = (r^2 + l_i^2 - 2l_ir \cos \theta_i)^{1/2}$$

and

$$\Delta\phi = \sum \frac{e_i}{r_i} \simeq \left(\sum e_i\right)\frac{1}{r} + \frac{1}{r^2}\left(\sum e_il_i \cos \theta_i\right) + \cdots .$$

If z_i is the vector QP_i and j is a unit vector in the direction QP, then $\sum e_il_i \cos \theta_i = (\sum e_iz_i) \cdot j = P \cdot j$ for the polarization $P = \sum e_iz_i$. Thus, we have a contribution to the potential in the form

$$\phi' = \iiint \frac{\rho}{r} dV + \iiint \frac{1}{r^2}(P \cdot j) dV + \cdots .$$

(a) What is the form of the higher terms in the expression for $\Delta\phi$?

(b) How does polarization contribute to potential on the surface of a body? Discuss the ambiguity in the expression for ϕ when both volumetric and surface terms are used.

(c) Suppose the charges occurred in matched pairs, e_i, $-e_i$, with displacement vector z between them. What is the formula for the contribution of such "dipoles" to the potential?

6.19. Prove the Gauss result for a repulsive charge distribution.

6.20. The argument in the text used to derive Euler's equation in the calculus of variations assumes that one can differentiate $(\partial F/\partial y')(x, y, y')$ and thus that the second derivative of the unknown function exists. On the other hand, if y and y' are continuous, then usually we can integrate $\partial F/\partial y$. Let $G(x) = \int_0^x (\partial F/\partial y)dx + C$. Then integration by parts yields

$$0 = \int_0^{x_0}\left(\frac{\partial F}{\partial y}\delta y + \frac{\partial F}{\partial y'}\delta y'\right)dx = \int_0^{x_0}\left[\frac{\partial F}{\partial y'} - G(x)\right]\delta y'\, dx + G\,\delta y\Big|_0^{x_0} .$$

Now $\delta y = 0$ at $x = 0$ and $x = x_0$ and thus $\int_0^{x_0} \delta y' \, dx = 0$ for all permissible variations δy. We have, then, instead of Euler's equation the result that

$$\frac{\partial F}{\partial y'}(x, y, y') = \int_0^{x_0} \frac{\partial F}{\partial y} \, dx + k$$

for some constant k. What are sufficient conditions that insure that y' has a continuous derivative and under what conditions does y' have a right and left derivative?

6.21. When one can as a practical matter assume the existence of a potential energy function of configuration, the Lagrangian variation theory can be applied to a nonrigid body. Consider, for example, a vibrating stretched elastic string satisfying Hooke's law. Consider a piece of string of length λ_0 when it is subject to no tension. If this piece is stretched to the length $\lambda = \lambda_0 + \mu$, then Hooke's law states that there is a constant h such that the tension required is $F = h\mu/\lambda_0$. The work done in stretching this piece is $W = h\mu^2/2\lambda_0 = h(\lambda - \lambda_0)^2/2\lambda_0$.

Suppose the original string has unstretched length l_0 and is stretched to a length l and laid along the x axis from $(0, 0)$ to $(l, 0)$. A vibration of this stretched string can be described by two functions $u(x, t)$, $v(x, t)$ with $0 \leqslant x \leqslant l$, $0 \leqslant t$, which are such that the point $(x, 0)$ on this initially stretched string is at the point $(x + u, v)$ at the time t. We assume that $u(x, 0)$, $v(x, 0)$, $\partial u(x, 0)/\partial t$, $\partial v(x, 0)/\partial t$ are given and that the endpoints are held fixed; i.e., $u(0, t) = v(0, t) = 0$, $u(l, t) = v(l, t) = 0$ for $t > 0$. The energy of configuration is obtained by considering the piece of string that in the initial stretched state lies between $(x, 0)$ and $(x + dx, 0)$. This has unstretched length $l_0 \, dx/l$ and at time t has length $[(1 + u_x)^2 + v_x^2]^{1/2} \, dx$. We assume that the radical can be approximated by means of the formula $(1 + a)^{1/2} = 1 + a/2$, since u_x and v_x are small. Then the configuration potential energy is

$$W = h \frac{l - l_0}{2l_0} \int_0^l [(1 + u_x)^2 + v_x^2 - 1] dx + K$$

for a constant K. If M is the total mass of the string, the kinetic energy is

$$T = \frac{M}{2} \int_0^l (v_t^2 + u_t^2) dx.$$

If $c^2 = h(l - l_0)/Ml_0$, the variation principle becomes

$$0 = \delta \int_{t'}^{t''} \int_0^l \{v_t^2 + u_t^2 - c^2[(1 + u_x)^2 + v_x^2]\} dx \, dt,$$

and the Euler equations are

$$c^2 u_{xx} = u_{tt} \quad \text{and} \quad c^2 v_{xx} = v_{tt}.$$

The solutions must satisfy the boundary and initial conditions given above.

6.22. Obtain the Euler equation for the brachistochrone and integrate it into the given parametric form. What is the geometric interpretation of the name "cycloid"? How can one determine a so that the curve goes through (x_0, y_0)?

6.23. Consider oxygen gas at atmospheric pressure and $0°C$ temperature. Show that the number of impacts of molecules during a microsecond on a square of side 1 μm on the container wall is about 2 billion. (Use the method in Chapter 1 of Mayer and Mayer.[17])

6.24. Show that if a one-to-one mapping of the basic point set of a manifold takes one atlas of a collection into another atlas of the same collection it takes every atlas of the collection into one of the collection.

6.25. Show that two changes of charts, say, from x coordinates to y coordinates and then from y to z yield the same relation between the Γ^i_{jk} associated with x and the Γ''^i_{jk} associated with the z as a direct change from x to z.

6.26. Determine the integrability conditions for the system

$$\frac{\partial y^i}{\partial x^\alpha}\Gamma^\alpha_{jk} + \frac{\partial^2 y^i}{\partial x^j\,\partial x^k} = 0.$$

6.27. Consider the equation

$$\frac{\partial g_{ij}}{\partial x^k} + g_{\alpha j}\Gamma^\alpha_{ki} + g_{i\alpha}\Gamma^\alpha_{kj} = 0$$

for determining the Γ^i_{jk} under the assumption that the g_{ij} are symmetric but without assuming symmetry for the Γ^i_{jk}. Show that under these circumstances the Γ^α_{ij} must be symmetric in the subscripts. Discuss the situation in which the g_{ij} are antisymmetric.

References

1. Abraham, Max, and Becker, Richard, *Electricity and Magnetism* (English translation), Hafner Publishing Company, New York (1932).
2. Bergmann, P. G., *Mechanics and Electrodynamics* (reprint), Dover Publications, Inc., New York (1962).
3. Bliss, G. A., *Lectures on the Calculus of Variations*, University of Chicago Press, Chicago (1946).
4. Boyer, Carl B., *The History of the Calculus and Its Conceptual Development* (2nd edition), Dover Publications, Inc., New York (1959).
5. Brillouin, Leon, Les Tenseurs en Mechanique et en Elasticité (reprint), Dover Publications, New York (1946).
6. Cajori, Florian, *Sir Isaac Newton's Mathematical Principles of Natural Philosophy and His System of the World* (Andrew Motte translation), University of California Press, Berkeley, California (1946), pp. 29ff.
7. Eisenhart, L. P., *Riemannian Geometry* (reprint), Princeton University Press, Princeton, New Jersey (1950).
8. Flanders, Harley, *Differential Forms*, Academic Press, New York (1963).
9. Heath, Sir Thomas, *A History of Greek Mathematics*. Vol. II, Oxford at the Clarendon Press, Oxford, England (1921).
10. Heitler, W., *The Quantum Theory of Radiation* (3rd edition), Clarendon Press, Oxford, England (1954).
11. Helgason, Sigurdur, *Differential Geometry and Symmetric Spaces*, Academic Press, New York (1962).
12. Kellogg, O. D., *Foundations of Potential Theory* (reprint), Dover Publications, Inc., New York (1953).
13. Klein, Jacob, *Greek Mathematical Thought and the Origin of Algebra* (English translation), The M.I.T. Press, Cambridge, Massachusetts (1968).
14. Lamb, Sir Horace, *Hydrodynamics* (reprint), Dover Publications, Inc., New York (1945).
15. Lanczos, Cornelius, *The Variational Principles of Mechanics* (3rd edition), University of Toronto Press, Toronto (1966).
16. Mason, Max, and Weaver, Warren, *The Electromagnetic Field*, Dover Publications, Inc., New York.

17. Mayer, J. E., and Mayer, M. P., *Statistical Mechanics*, John Wiley and Sons, Inc., New York (1966).
18. Robinson, Abraham, Non-Standard Analysis, North-Holland Publishing Co., Amsterdam (1966).
19. Smith, David Eugene, *A Source Book in Mathematics*, McGraw-Hill Book Company, Inc., New York and London (1929).
20. Van der Waerden, B. L., *Science Awakening* (English translation), P. Noordhoff Ltd., Gröningen, Holland (1954).
21. Weyl, Herman, *Space, Time and Matter*, Dover Publications, Inc., New York (1922).
22. Whitney, Hassler, *Geometric Integration Theory*, Princeton University Press, Princeton, New Jersey (1957).
23. Whittaker, E. T., *A Treatise on the Analytical Dynamics of Particles and Rigid Bodies*, Dover Publications, Inc., New York (1944).
24. Whittaker, Sir Edmund, *A History of the Theories of Aether and Electricity*, Harper Brothers, New York (1960).

7

Energy

7.1. The Motion of Bodies

In discussing the flight trainer we obtained a description of the motion of a rigid body in terms of the external forces acting on it. In general, however, bodies are not so rigid that relative motion between the parts can be completely neglected. In many cases the general shape of the body is still retained, so that it seems desirable to distinguish a gross motion for the body as a whole and a relative motion for the parts. The distinction we will make between the two motions does have certain arbitrary aspects, but these are consequent on practical exigencies.

Newton's second law that action and reaction are equal has certain consequences for an arbitrary system of N particles. Let the coordinates of these particles be y_j^i, $i = 1, 2, 3, j = 1, ..., N$ and let y_j denote the vector displacement of the jth particle, and F_j the resultant force on the jth particle. In a Galilean frame we have $m_j \ddot{y}_j = F_j$ and consequently for the center of gravity $Y = (\Sigma m_\alpha y_\alpha)/M$, we have

$$M \ddot{Y} = \sum_\alpha F_\alpha = F_R . \tag{1}$$

In the summation, F_R, forces between particles cancel by Newton's second law and we need only sum over forces on the particles due to external aspects.

We also have $m_j(\ddot{y} - \ddot{Y}) = F_j - m_j F_R/M$ and

$$m_j(y_j - Y) \times (\ddot{y}_j - \ddot{Y}) = (y_j - Y) \times (F_j - m_j F_R/M),$$

which can be written

$$\frac{d}{dt} [m_j(y_j - Y) \times (\dot{y}_j - \dot{Y})] = (y_j - Y) \times F_j - (y_j - Y)m_j F_R/M.$$

165

If we sum over the particles, the last term contributes zero so that we have for $L = \sum m_\alpha(y_\alpha - Y) \times (\dot{y}_\alpha - \dot{Y})$

$$\frac{dL}{dt} = \sum_\alpha (y_\alpha - Y) \times F_\alpha = T. \tag{2}$$

In the summation for T, the collinearity of action and reaction implies that again we need only add over forces on particles that have an external effect.

When the relation (1) is applied to substance in general it shows that if the motion is not disruptive it must be possible to introduce an intensive quantity f, the "force per unit volume" corresponding to Newton's third law. Let $\mathfrak{A}$ be any region in the substance. The center of gravity of the material in $\mathfrak{A}$ is given by

$$\left(\iiint_{\mathfrak{A}} \rho \, dV \right) Y = \iiint_{\mathfrak{A}} y\rho \, dV.$$

$\dot{Y}$ corresponds to the motion of the material in $\mathfrak{A}$. We can regard the integrals $\iiint \rho \, dV$ and $\iiint y\rho \, dV$ as summations relative to a large number of small particles with $dM = \rho \, dV$ and thus

$$\frac{D}{Dt} \iiint \rho \, dV = 0, \qquad \frac{D}{Dt} \iiint y\rho \, dV = \iiint v\rho \, dV.$$

This can also be expressed by the relation

$$\frac{D}{Dt}(\rho \, dV) = \left(\frac{D\rho}{Dt} \right) dV + \rho \frac{D(dV)}{Dt} = \left(\frac{D\rho}{Dt} + \rho \operatorname{div} v \right) dV$$

$$= \left[\frac{\partial \rho}{\partial t} + \frac{\partial(\rho v^1)}{\partial x^1} + \frac{\partial(\rho v^2)}{\partial x^2} + \frac{\partial(\rho v^3)}{\partial x^3} \right] dV = 0$$

by the equation of continuity. Thus, we have

$$M_{\mathfrak{A}} \dot{Y} = \iiint_{\mathfrak{A}} \rho v \, dV \quad \text{and} \quad M_{\mathfrak{A}} \ddot{Y} = \iiint_{\mathfrak{A}} \rho \ddot{y} \, dV.$$

The force acting on the substance in $\mathfrak{A}$ is an extensive quantity with value $M_{\mathfrak{A}} \ddot{Y}$. The extensive nature can be obtained by considering the force on contiguous regions or just from the above formula for $M \ddot{Y}$. It is then the integral of an intensive function f such that

$$\iiint_{\mathfrak{A}} f \, dV = \iiint_{\mathfrak{A}} \rho \ddot{y} \, dV. \tag{3}$$

Thus, $f = \rho \ddot{y}$. We will return to this definition of f as an intensive function. We must also consider the precise significance of $\ddot{y}$.

With this definition of f we again have $F_R = \iiint f \, dV$ and $M\ddot{Y} = F_R$ for a body or substance of finite extent. We can consider Newton's third law as expressed by the equation $f \, dV = \rho \ddot{y} \, dV$ and we have $(f - F_R \rho/M) dV = \rho(\ddot{y} - \ddot{Y}) dV$ and

$$dV (y - Y) \times (f - F_R \rho/M) = \frac{d}{dt} [\rho(y - Y) \times (\dot{y} - \dot{Y}) \, dV].$$

Thus, summing over the space that contains the substance yields

$$T = \iiint (y - Y) \times f \, dV = \frac{d}{dt} L = \frac{d}{dt} \iiint \rho[(y - Y) \times (\dot{y} - \dot{Y})] dV. \quad (4)$$

If we sum only over a region $\mathfrak{A}$, then Y is to be taken as center of gravity $Y_\mathfrak{A}$ for the substance in $\mathfrak{A}$ and the result is

$$T_\mathfrak{A} = \iiint_\mathfrak{A} [(y - Y_\mathfrak{A}) \times f] dV = \frac{d}{dt} \iiint \rho[(y - Y_\mathfrak{A}) \times (\dot{y} - \dot{Y}_\mathfrak{A})] dV.$$

The quantity to be differentiated is $L_\mathfrak{A}$, the angular momentum of the substance in $\mathfrak{A}$.

We can now introduce a Cartesian frame centered at Y by means of an orthogonal matrix i

$$y = Y + ix,$$

where x is the position vector in the new frame. Let Ω be the spin vector determined by the relation $di/dt = i(\Omega \times)$. Then

$$L = \iiint \rho[(y - Y) \times (\dot{y} - \dot{Y})] dV = \iiint \rho\{ix \times (\dot{x} + \Omega \times x)\} dV$$

$$= i\left\{\iiint \rho(x \times \dot{x}) dV + \iiint \rho[x \times (\Omega \times x)] dV\right\}$$

$$= i\left\{l + \iiint \rho[x \times (\Omega \times x)] dV\right\}.$$

Here $l = \iint \rho(x \times \dot{x}) dV$ is the angular momentum in the new coordinate system. The second term can be expressed as a matrix transformation on the vector Ω, i.e.,

$$\iiint \rho[x \times (\Omega \times x)] dV = J\Omega,$$

where the matrix J has elements

$$J_{ij} = -\iiint \rho x^i x^j \, dV + \delta_{ij} \iiint \rho x \cdot x \, dV$$

and

$$L = i(l + J\Omega).$$

If we knew how to calculate f, we have the differential equation $m\ddot{Y} = F_R$ for Y. However, in the general case there is an ambiguity relative to Ω and the motion of the configuration as expressed in terms of x so that one does not have a practical way to determine Ω and i.

But to reap any advantage from the introduction of the orthogonal matrix i, we must suppose that we are dealing with a body or object in which the shape is essentially preserved. We have seen that in the case where the shape is actually preserved, that is, in the case of a rigid body, we can introduce an orthogonal matrix i so that if y describes the motion of a point on the body then x is a constant. Furthermore ρ is a function of x. We will interpret the statement that "the shape is essentially preserved" as corresponding to the fact that there exists an orthogonal matrix i for which the motion of point is expressed by an equation $x = x_0 + u$, where x_0 is a constant vector and u is small. We suppose that there is a reference situation in which $u = 0$ and that $\rho_0 = \rho_0(x_0^1, x_0^2, x_0^3)$ describes the density in this reference situation.

The motion then is described by giving Y, i, and u as functions of x_0^1, x_0^2, x_0^3. In the partition into particles, we use the reference situation. Thus we have

$$\iiint \rho_0 x_0 \, dV_0 = 0 \quad \text{and} \quad \iiint \rho_0 u \, dV_0 = 0.$$

We assume that there is a vector function f_0 such that Newton's third law can be expressed $\rho_0 \ddot{y} \, dV_0 = f_0 \, dV_0$. This yields

$$M\ddot{Y} = \iiint f_0 \, dV_0 = F_R,$$

where of course the contributions to f_0 of inner forces can be ignored, and similarly we have

$$T = \iiint [(y - Y) \times f_0] dV_0 = \frac{d}{dt} \iiint \rho_0 [(y - Y) \times (\dot{y} - \dot{Y})] dV_0.$$

Now we can use $y - Y = i(x_0 + u)$, $\dot{y} - \dot{Y} = i[\dot{u} + \Omega \times (x_0 + u)]$ and obtain

$$L = \iiint \rho_0 [(y - Y) \times (\dot{y} - \dot{Y})] dV_0$$

$$= i \iiint \rho_0 [(x_0 + u) \times (\Omega \times x_0 + \dot{u} + \Omega \times u)] dV_0$$

$$= i \left[\iiint \rho_0 [x_0 \times (\Omega \times x_0)] dV_0 + \mathscr{E}(u, x_0) \right] = i(J_0\Omega + \mathscr{E}),$$

where J_0 is calculated from the reference position, i.e.,

$$J_{0ij} = -\iiint \rho_0 x_0^i x_0^j \, dV_0 + \delta_{ij} \iiint \rho x_0 \cdot x_0 \, dV_0.$$

We can represent f_0 in the x frame by letting $f_0 = ig_0$. The torque relation then becomes

$$S_0 = \iiint (x_0 + u) \times g_0 \, dV = \left(\frac{d}{dt} + \Omega x\right)(J_0 \Omega + \mathscr{E}).$$

This suggests that the x frame be specified by the conditions $M\ddot{Y} = F_R$ and $S_0 = (d/dt + \Omega \times)J_0\Omega$ and that u be determined by the relation

$$g_0 = \rho_0 \left\{ G_R/M + \left(\frac{d}{dt} + \Omega \times\right)[\ddot{u} + \Omega \times (x_0 + u)]\right\},$$

where $F_R = iG_R$. The expression for u may appear complicated, but it is just the form Newton's third law takes in a moving coordinate system. In many cases g_0 will depend on u and its partial derivatives relative to x_0^1, x_0^2, x_0^3 so that the u equation can be regarded as a second-order partial differential equation on $u(x_0^1, x_0^2, x_0^3, t)$.

It is not clear mathematically that the functions u obtained will be small. If we multiply the above equation by i and integrate relative to dV_0, we will have

$$0 = i\left(\frac{d}{dt} + \Omega \times\right)\left(\frac{d}{dt} + \Omega \times\right)\iiint \rho_0(x_0 + u)dV_0 = \frac{d^2}{dt^2}\left(i\iiint \rho_0 u \, dV_0\right).$$

$\iiint f_0 \, dV = F_R$, ρ_0 does not depend on t, and $\iiint \rho_0 x_0 \, dV_0 = 0$. This yields

$$i\iiint \rho_0 u \, dV_0 = at + b.$$

We can suppose that auxiliary conditions are imposed on u so that the vectors a and b are zero and thus $\iiint \rho_0 u \, dV_0 = 0$. This is consistent with u small but does not imply it. Nevertheless the critical question is to determine f_0.

7.2. The Stress Tensor

Let us now consider the general case of a substance in a region with a Cartesian frame from a Galilean system. We wish to specify a vector function f of the variables y^1, y^2, y^3, and t such that

$$\iiint_{\mathfrak{A}} f \, dV = F_R = \iiint_{\mathfrak{A}} m\ddot{y} \, dV$$

is the resultant force on the substance in the region $\mathfrak{A}$ for any $\mathfrak{A}$ that contains the substance.

F_R is composed of two types of forces. We have, for example, forces that are distributed through the bulk of the material and that can be expressed directly as integrals,

$$\Phi = \iiint \phi(y) \, dV.$$

An example is gravity. For a small region near the earth this can be expressed in terms of a constant vector g as $g \iiint \rho \, dV$. In general, however, the force due to gravity can be expressed as

$$\iiint \rho g \, dV,$$

where g may depend on y^1, y^2, y^3, and t.

Another force that contributes to F_R is the resultant of the forces that the rest of the substance exerts on the stuff in $\mathfrak{A}$ through the boundary $\mathfrak{B}$. If we can find a suitable integral expression for this force we will be able to apply Gauss' theorem to obtain a volume integral.

Consider an infinitesimal element $d\sigma$ of surface with unit normal (n_1, n_2, n_3) determining an upper side for $d\sigma$. Our notion of contact indicates that the substance on the lower side of $d\sigma$ presses on the substance on the upper side with a force F, which is proportional to the area of $d\sigma$ and depends on the position y^1, y^2, y^3, t and the components n_1, n_2, n_3 of the normal, i.e., $F = F(n, y) \, dA$.

There is a standard argument that we will sketch that shows that F is linear in the n_1, n_2, n_3, i.e., that

$$F = dA i_\alpha a_{\alpha\beta} n_\beta$$

and $a_{ij} = a_{ji}$. In general the a_{ij} will be functions of y^1, y^2, y^3, and t.

Consider a point P_0, y_0^1, y_0^2, y_0^3, and let a normal direction (n_1, n_2, n_3) be chosen. For convenience we will assume $n_i > 0$. Let p be a small positive number and consider the plane l with equation

$$n_1 x^1 + n_2 x^2 + n_3 x^3 = p$$

in a coordinate system with origin at P_0 and axes parallel to the original. Let P_i, $i = 1, 2, 3$, denote the intercept of l with the x^i axis. The intercept values are $x^i = p/n_i$. The volume of the tetrahedron, $P_0 P_1 P_2 P_3$, is $p^3/6n_1 n_2 n_3$ and the area, A, of the triangle $P_1 P_2 P_3$ in the plane l is $p^2/2n_1 n_2 n_3$. The triangle $P_0 P_2 P_3$ in the x^2, x^3 plane has area $A_1 = p^2/2n_2 n_3$. Similarly we can define A_2 and A_3 as the areas of the triangles $P_0 P_3 P_1$ and $P_0 P_1 P_2$, respectively. Thus, $A_i = A n_i$.

Consider now the forces on the various faces of the tetrahedron $P_0P_1P_2P_3$ by the exterior substance. That on $P_1P_2P_3$ is $-F(n)A$. Now i_j is the unit normal to the face with area A_j and is directed into the tetrahedron. The corresponding force on this face is $F(i_j)A_j = F(i_j)n_jA$. Thus, the resultant of the forces on the faces of the tetrahedron is $[F(i_\beta)n_\beta - F(n)]A$.

Now let $\Phi = \phi V$ be the resultant of the body forces on the tetrahedron, where $V = pA/3$ is the volume. Let X denote the position vector for the center of gravity of the tetrahedron. Then

$$[F(i_\beta)n_\beta - F(n)]A = Vp\ddot{X} - \phi V = pA(\rho\ddot{X} - \phi)/3 .$$

If we cancel A and let p approach zero, we obtain $F(n) = F(i_\beta)n_\beta = i_\alpha a_{\alpha\beta}n_\beta$ when $a_{\alpha j}$ is defined by $F(i_j) = i_\alpha a_{\alpha j}$.

Thus, $F(n)$ is related to the normal vector n by the transformation $F(n) = an$, where a is the matrix (a_{ij}). The symmetry of the matrix a is established by considering the torque equation for a small cube of side a of the substance centered at P_0. The forces on opposing faces of the cube constitute couples that contribute a total torque of

$$-a^3i_\alpha \times F(i_\alpha) = a^3(a_{23} - a_{32}, a_{31} - a_{13}, a_{12} - a_{21}).$$

The angular momentum of the cube corresponding to a spin vector Ω is

$$L_c = \int_{-a/2}^{a/2} \int_{-a/2}^{a/2} \int_{-a/2}^{a/2} \rho(P_0)[x \times (\Omega \times x)]dx^1\, dx^2\, dx^3$$

$$= (\rho_0 a^5/6)\Omega.$$

We suppose that the body forces are continuously differentiable; i.e., $\phi(x^1, x^2, x^3) = \phi(P_0) + (\partial\phi/\partial x^\alpha)\, x^\alpha$. The torque T_ϕ due to the body forces is then

$$T_\phi = \iiint x \times \phi\, dx^1\, dx^2\, dx^3 = (\rho_0 a^5/12)\,\text{curl}\,\phi.$$

An argument similar to the previous one now shows that $a_{ij} = a_{ji}$.

The symmetry of the matrix a is of considerable significance. Suppose at the point P_0 we replace the x^1, x^2, x^3 coordinate system with another Cartesian coordinate system $x^{1'}$, $x^{2'}$, $x^{3'}$ with unit vectors i'_1, i'_2, i'_3 along the axes instead of i_1, i_2, i_3. This determines an orthogonal matrix $j = (j_{rs})$ by the relation $i'_r = j_{r\beta}i_\beta$ or equivalently $i_s = i'_\alpha j_{\alpha s}$. We consider the coordinates n^1, n^2, n^3 of n in the x system as constituting a one-column matrix, which we also denote by n and similarly for n'. The relation $i'_\alpha n^{\alpha'} = n = i_\beta n^\beta = i'_\alpha j_{\alpha\beta}n^\beta$ yields $n^{r'} = j_{r\beta}n^\beta$ or $n' = jn$ for the one-column matrices. Similarly we have for $F(n)$ and $F(n')$ as one-column matrices $F(n') = jF(n)$. This also indicates the transformation of the matrix a, since $a'n' = F(n') = jF(n) = jan = jaj^{-1}n'$ or $a' = jaj^{-1} = jaj'$.

Since a is symmetric, an orthogonal matrix j can be chosen so that a' is diagonal. Now if n is a given direction, $F(n)$ perpendicular to the surface element $d\sigma$ is equivalent to $F(n) = \lambda n = an$, or n is a characteristic vector of the matrix a. Since a is symmetric there always are three such directions at each point y^1, y^2, y^3 that are mutually orthogonal, and these can be taken as i'_1, i'_2, and i'_3 to determine the matrix j. If the characteristic values of a which we denote by λ_1, λ_2, λ_3 are distinct, these three directions are determined up to a constant at each point y^1, y^2, y^3. If the λ_i remain distinct through out a region $\mathfrak{A}$ in the substance, we can consider them as varying continuously and the vectors i'_1, i'_2, and i'_3 as well.

The matrix a is a "tensor" in the original sense of the word and gives the stress $F(n)$ throughout the substance. It is clear that we can associate a with a field of three mutually perpendicular vectors $\lambda_1 i'_1$, $\lambda_2 i'_2$, $\lambda_3 i'_3$ distributed through the substance such that $F(i'_r) = \lambda_r i'_r$, where λ_r is the force per unit area on an infinitesimal surface element $d\sigma$ perpendicular to the direction i'_r. Conversely, given such a field, the tensor a is determined.

The expression for the force on an infinitesimal element of surface permits us to express the resultant of the contact forces on the boundary $\mathfrak{B}$ of a region $\mathfrak{A}$ at a time t.

$$
\begin{aligned}
F_c &= -\iint_{\mathfrak{B}} i_\alpha a_{\alpha\beta} n_\beta \, dA \\
&= -\iint_{\mathfrak{B}} i_\alpha (a_{\alpha 1} \, dy^2 \, dy^3 + a_{\alpha 2} \, dy^3 \, dy^1 + a_{\alpha 3} \, dy^1 \, dy^2) \\
&= -\iiint_{\mathfrak{A}} i_\alpha \left(\frac{\partial a_{\alpha 1}}{\partial y^1} + \frac{\partial a_{\alpha 2}}{\partial y^2} + \frac{\partial a_{\alpha 3}}{\partial y^3} \right) dy^1 \, dy^2 \, dy^3.
\end{aligned}
$$

Here the a_{ij} are to be considered as functions of y^1, y^2, y^3, and t, that is, of the situation as it appears in the medium. Thus, the force acting on the substance in a region $\mathfrak{A}$ is given by an integral

$$
F_{\mathfrak{A}} = \iiint \left[\Phi(y^1, y^2, y^3) - i_\alpha \frac{\partial}{\partial y^\beta} a_{\alpha\beta} \right] dy^1 \, dy^2 \, dy^3,
$$

where Φ is a vector expression for the body forces.

The Eulerian description describes the activity of a substance in terms of spatial position and time. The movement of a body in particular would be expressed by its density $\rho = \rho(y^1, y^2, y^3, t)$ and three velocity components $v^i = v^i(y, t)$. For this description the acceleration $\ddot{y}$ is the intrinsic derivative of the velocity vector

$$
\ddot{y} = \frac{\partial v}{\partial t} + \frac{\partial v}{\partial y^\alpha} v^\alpha,
$$

and Newton's law $f\,dV=\rho\ddot{y}\,dV$ becomes

$$\Phi(y^1,\,y^2,\,y^3)-i_\alpha\frac{\partial}{\partial y\beta}\,a_{\alpha\beta}=\frac{\partial v}{\partial t}+\frac{\partial v}{\partial y^\alpha}\,v^\alpha.$$

With the equation of continuity we now have four equations on the four quantities $\rho,\,v^1,\,v^2,\,v^3$, provided we know the a_{ij} values.

In the alternate descriptions, points in the substance are identified by their position, say, $x_0^1,\,x_0^2,\,x_0^3$ in a certain reference situation, and the current position is given by $y=y(x_0,\,t)$. We have used this in describing bodies whose shape is approximately preserved. For a fixed value of t, this yields a spatial transformation $x_0\rightarrow y$. Let

$$J=\frac{\partial(y^1,\,y^2,\,y^3)}{\partial(x_0^1,\,x_0^2,\,x_0^3)}\,.$$

Under this transformation we can regard a y region $\mathfrak{A}$ and its boundary $\mathfrak{B}$ as the images of corresponding $\mathfrak{A}_0$ and $\mathfrak{B}_0$ in x_0 space. From the formula for changing variables in an integral we have $\rho(y,\,t)J=\rho(x_0)$. Similarly the resultant of the body forces is given by

$$\iiint_{\mathfrak{A}}\Phi(y)\,dV=\iiint_{\mathfrak{A}_0}\Phi(y(x_0,\,t))J\,dV_0.$$

If the expressions $\partial a_{i\alpha}/\partial y^\alpha$ are available as functions of y or even their partial derivatives, we can make a change to a dV_0 integral. Alternately, we can transform the integral

$$\iint_{\mathfrak{B}}i_\alpha(a_{\alpha 1}\,dy^2\,dy^3+a_{\alpha 2}\,dy^3\,dy^1+a_{\alpha 3}\,dy^1\,dy^2)$$

by using equations in the form

$$dy^2\,dy^3=\left(\frac{\partial y^2}{\partial x_0^\alpha}\,dx_0^\alpha\right)\left(\frac{\partial y^3}{\partial x_0^\alpha}\,dx_0^\alpha\right)=\eta_1^1\,dx_0^2\,dx_0^3+\eta_1^2\,dx_0^3\,dx_0^1+\eta_1^3\,dx_0^1\,dx_0^2,$$

where

$$\eta_1^1=\frac{\partial(y^2,\,y^3)}{\partial(x_0^2,\,x_0^3)}\,,\quad\text{etc.}$$

The resultant of the boundary contact force is then

$$-\iint_{\mathfrak{B}_0}i_\alpha(a_{\alpha\beta}\eta_1^\beta\,dx_0^2\,dx_0^3+a_{\alpha\beta}\eta_2^\beta\,dx_0^3\,dx_0^1+a_{\alpha\beta}\eta_3^\beta\,dx_0^1\,dx_0^2).$$

We can write $a_{ij}^0=a_{i\alpha}\eta_j^\alpha$ and consider it as depending on y and the partials of y

relative to the space variable. Of course, we must look into this further. The boundary force now can be expressed

$$-\iiint_{\mathfrak{A}_0} i_\alpha \frac{\partial}{\partial x^\beta} (a^0_{\alpha\beta})\, dV_0.$$

Newton's equations now become

$$\Phi(y)J - i_\alpha \frac{\partial}{\partial x^\beta_0}(a^0_{\alpha\beta}) = \rho(x_0)\frac{\partial^2 y}{\partial t^2}.$$

These are three equations on the three functions $y^1(x_0, t)$ $y^2(x_0, t)$, $y^3(x_0, t)$, and to use them we must find the a^0_{ij}. If we wish to consider motion in the form $y = Y + jx$ or $y = Y + j(x_0 + u)$, these, of course, become equations on $x = x(x_0, t)$ or $u = u(x_0, t)$.

7.3. Deformation and Stress

We must now discuss ways in which the tensor components a_{ij} can be determined. Stresses within a body are usually associated with a "change of shape." Stresses appear because the substance is distorted or strained. The notion of change of shape requires that we be able to identify points in the substance and determine their positional history. Also, there is a reference shape or position for the body that corresponds to an "undistorted" shape and that can be used to specify points in the body. Distortion is purely a spatial relation, i.e., if we take a snapshot of the body at any instant in time, we can determine the distortion by comparing purely spatial aspects with the reference.

It is convenient to assume that in the reference shape the center of gravity is at the origin. We chose a Cartesian coordinate system for this reference position and specify a point P_0 by coordinates x^1_0, x^2_0, x^3_0. Now, if we move the substance congruently so that P_0 goes into a point Q_0 with coordinates $y_0 = Y + jx_0$, with orthogonal matrix j, then the new position is also an undistorted reference shape. A distorted position can be considered as corresponding to a mapping $x_0 \rightarrow x(x_0)$ if P_0 goes into the point Q with coordinates $y = Y + jx$. The distortion then does not depend on the vector Y, and the distortion is the same for all mappings $x_0 = \bar{x}(x_0)$ for which $\bar{x}(x_0) = j'x(j''x_0)$ where j'' and j' are orthogonal matrices. In the actual motion Y can be taken as $Y(t)$ and we may be able to determine $j = j(t)$ so that $x(x_0, t) = x_0 + u(x_0, t)$ for u small. But distortion is not a notion associated with time but is a property of a class of spatial transformations. A transformation $x_0 \rightarrow x(x_0)$ determines a class of transformations $x_0 \rightarrow Y + j'x(j''x_0)$, but any transformation in this class also determines the same class in this way.

Consider then the mapping $x_0 \to x(x_0)$, which is given by three functions, $x^i = x^i(x_0^1, x_0^2, x_0^3)$. We will suppose that these functions are continuously differentiable up to the second order, that the Jacobian

$$K = \frac{\partial(x^1, x^2, x^3)}{\partial(x_0^1, x_0^2, x_0^3)}$$

is positive, and that it is bounded away from zero. Thus, the amount of volume compression at a point is restricted.

A point P in the neighborhood of P_0 will have coordinates $x_0^i + l^i \, dr$ with dr small and $(l^1)^2 + (l^2)^2 + (l^3)^2 = 1$. Under the transformation $x_0 \to x(x_0)$, P will go into a point Q with coordinates $x^i + J_\alpha^i l^\alpha \, dr$ with $J_j^i = \partial x^i / \partial x_0^j$. For dr fixed, we have, then, a transformation J such that the unit vector l is taken into $Jl = (J_\alpha^1 l^\alpha, J_\alpha^2 l^\alpha, J_\alpha^3 l^\alpha)$. Correspondingly we have a quadratic form $\alpha_{\alpha\beta} l^\alpha l^\beta = Jl \cdot Jl$ for the transformed length squared of l with positive definite matrix α. We can find a positive definite matrix A such that $A^2 = \alpha$. A has the same characteristic vectors as α, and the characteristic values of A are the square roots of the corresponding values of α. Thus $Al \cdot Al = A^2 l \cdot l = \alpha_{\alpha\beta} l^\alpha l^\beta = Jl \cdot Jl = J'Jl \cdot l$. We also have $A^2 = \alpha = J'J$, and for any vector l, $\|Al\| = \|Jl\|$; i.e., the length of Al equals that of Jl.

The matrix A is nonsingular, since $Al = 0$ implies $Jl = 0$ and $l = 0$. The matrix $j = JA^{-1}$ has the property that for any vector l,

$$\|jl\| = \|J(A^{-1}l)\| = \|A(A^{-1}l)\| = \|l\|.$$

This yields

$$jl \cdot jl' = [\|j(l + l')\|^2 - \|j(l - l')\|^2]/4 = (\|l + l'\|^2 - \|l - l'\|^2)/4 = l \cdot l',$$

and hence j is orthogonal. We have $jA = J$. This is the "canonical resolution" for J with A positive definite and j orthogonal. We have an alternate resolution for J in the form $J = Bj$ with $B = jAj^{-1}$ also positive definite. Corresponding to $J'J = A^2$, we have, since $J' = j'B = j^{-1}B$, $JJ' = B^2$.

Thus, the transformation $x_0 \to x(x_0)$ determines a field of transformations, $J = (\partial x^i / \partial x_0^j)$, as well as two other fields of matrices A and j. The matrix A can be interpreted as follows. Take a small cube of the substance with vertex at P_0 and sides parallel to the three characteristic vectors of A. Let μ_1, μ_2, μ_3 be the corresponding characteristic values. Compress each side of the cube in the ratio of the corresponding μ_i. This should be done so that any smaller cube inside the given cube and with the same side directions is similarly compressed. The action of J, then, consists of such a compression followed by a rotation j. If we use the alternate resolution $J = Bj$, we first rotate the chosen cube by j and then compress as indicated in the given ratios, which are also the characteristic values of B. If we use any other transformation of the same class, J is replaced by $j'Jj'' = j'jAj'' = (j'jj'')j''^{-1}Aj''$.

Thus, the new vector field J' has $A' = j''^{-1}Aj''$. This is just a rotation of A by j'', which is the same at all points.

This indicates that either the compression due to A or that due to B corresponds to the distortion of the substance. In fact it can be shown that the field A determines the field j up to a constant multiplicative j_0.

If we consider the B compression of the rotated cube as associated with forces perpendicular to the faces of the cube, the area average of these forces yields a system of stresses of the type we have associated with the matrix a in the previous section. For now we have three perpendicular directions, and for each of these directions the stress on the plane surface to which it is normal is in the same direction. Each such direction is therefore a characteristic vector of a. We must still relate the face stresses, that is, the characteristic values λ_1, λ_2, λ_3 of a, with the compression ratios μ_1, μ_2, μ_3, which are the characteristic values of B.

We can now consider how this relationship can be determined experimentally with no reference to infinitesimals. For suppose a cube of side a of the undeformed material is confined within a very strong box of sides $\mu_1 a$, $\mu_2 a$, $\mu_3 a$ in such a way that the compression is similar throughout the substance. The stress on each face should be uniform and the stresses on opposing faces must be negatives of each other if the box does not move. We assume by symmetry that these stresses on opposing faces produce no torque, and hence these stresses are perpendicular to the faces. These arguments apply to any smaller cube within the compressed substance by the assumption of similar compression. Now if λ_1, λ_2, λ_3 are the values of the stresses on opposing faces, then one has a uniform situation throughout the substance in which the compression of the matrix B is associated with the tensor a with the same characteristic directions and characteristic values $\lambda_1, \lambda_2, \lambda_3$. If we specify the μ_i and determine the λ_i by measurements, we obtain λ_i as a function of μ_1, μ_2, μ_3, or we could proceed in the other direction. Referring the infinitesimal situation to a finite one is, of course, the essential idea of calculus, that is, "The Method of Fluxions."

Unfortunately there is a difficulty we must now face. In general when the λ_i are measured, they will be found to depend on the way the desired compression was obtained, that is, on the time history. If we measure very quickly the substance may get warm and yield, say, higher values than if we do it more slowly. If we compress it to higher values and then permit it to recede to the desired compression, the values of λ_1, λ_2, and λ_3 are often much lower than the ones obtained from a monotonic compression. Thus, the time history of the compression must also be specified for the experiment, and the history of the compression of the infinitesimal cube in the analysis must also be specified.

There are, however, certain practical cases in which simpler relations between the λ_i and μ_i can be assumed, at least as approximations. Let us consider the significance of the assumption that the λ_i are functions of the μ_i. Consider, then, a specific procedure for compressing the given cube of material from, say, $\mu_i = 1$ to values $\mu_1^0, \mu_2^0, \mu_3^0$. This means that each μ_i is given as a function of time, and we will suppose the procedure occurs in the time interval $0 \leqslant t \leqslant 1$. The work done in compressing the cube of material is then given by

$$W = a^3 \int_0^1 \left(-\lambda_1 \mu_2 \mu_3 \frac{d\mu_1}{dt} - \lambda_2 \mu_3 \mu_1 \frac{d\mu_2}{dt} - \lambda_3 \mu_1 \mu_2 \frac{d\mu_3}{dt} \right) dt,$$

where the λ_i are functions of the μ_i and hence of the time. But if we reverse this procedure the cube will now do precisely the same amount of work in decompressing, since the λ_i are determined by the μ_i. It follows, then, that the work W must be present in the form of potential elastic energy in the compressed cube. Thus, the assumption that the λ_i are functions of μ is equivalent to the existence of an elastic potential energy function u_e of μ_1, μ_2, μ_3 such that a cube of side a has, when compressed, the energy $a^3 u_e$. u_e is such that

$$du_e = -\lambda_1 \mu_2 \mu_3 \, d\mu_1 - \lambda_2 \mu_3 \mu_1 \, d\mu_2 - \lambda_3 \mu_1 \mu_2 \, d\mu_3,$$

and we have

$$\lambda_i = -\mu_i \frac{\partial u_e / \partial u_i}{\mu_1 \mu_2 \mu_3}.$$

For a substance for which such a u_e exists, the total potential energy is obtained by a summation in which a^3 is replaced by dV_0, the uncompressed reference volume. The extensive elastic potential energy is given by

$$U_e = \iiint u_e(\mu_1, \mu_2, \mu_3) \, dV_0.$$

There are circumstances in which the assumption of the existence of an elastic potential energy is reasonable. These include the possibility of small rapid oscillations in which heat flow can be either neglected or compensated for by dissipation terms. One also notices that u_e is ultimately dependent on certain functions of the partial derivatives, $\partial x^i / \partial x_0^j$, since the μ_i were obtained as follows. One forms the characteristic equation of the matrix $JJ' = B^2$, which has three coefficients, which are polynomials in the partials $\partial x^i / \partial x_0^j$. One solves for the roots of this equation and extracts the square roots.

Let us consider, then, the motion of a substance for which an elastic potential energy function exists. We suppose that its point P_0, which is at x_0 in the reference position, moves in accordance with

$$y = Y(t) + j(t)x(x^0, t).$$

For simplicity we will assume that Y, j and its spin vector Ω are given functions of t. The elastic potential energy is given by an integral

$$U_e = \iiint u_e \, dV_0.$$

We consider u_e to be a function of the $\partial x^i / \partial x^j_0$ and let u_{eij} denote the partial of u_e relative to $\partial x^i / \partial x^j_0$. We suppose that the potential energy of the body forces are given also by a spatial integral

$$V = \iiint \phi(x^1, x^2, x^3) \frac{\partial(x^1, x^2, x^3)}{\partial(x^1_0, x^2_0, x^3_0)} \, dV_0.$$

If $Y(t)$ is the center of gravity, the kinetic energy can be written as

$$T = \frac{1}{2} M \dot{Y}^2 + \frac{1}{2} \iiint \rho \left\| \frac{\partial x}{\partial t} + \Omega \times x \right\|^2 dV_0$$

$$= \frac{1}{2} M \dot{Y}^2 + \frac{1}{2} \iiint \rho \left(\frac{\partial x^1}{\partial t} + \Omega^2 x^3 - \Omega^3 x^2 \right)^2 dV_0$$

$$+ \frac{1}{2} \iiint \rho \left(\frac{\partial x^2}{\partial t} + \Omega^3 x^1 - \Omega^1 x^3 \right) dV_0$$

$$+ \frac{1}{2} \iiint \rho \left(\frac{\partial x^3}{\partial t} + \Omega^1 x^2 - \Omega^2 x^1 \right)^2 dV_0.$$

The Lagrangian, then, is $T - U_e - V$ and one must have a stationary value for $\int_{t'}^{t''} (T - U_e - V) dt$ relative to variations in the functions $x^i(x_0, t)$. The usual variation procedures then yield three partial differential equations, the first of which is the following

$$-\frac{1}{2} \rho \frac{\partial}{\partial t} \left(\frac{\partial x^1}{\partial t} + \Omega^2 x^3 - \Omega^3 x^2 \right) + \sum_\alpha \frac{\partial}{\partial x^\alpha} (u_{e'1\alpha} + \phi K^1_\alpha)$$

$$+ \rho \left[\Omega^3 \left(\frac{\partial x^2}{\partial t} + \Omega^3 x^1 - \Omega^1 x^3 \right) - \Omega^2 \left(\frac{\partial x^3}{\partial t} + \Omega^1 x^2 - \Omega^2 x^1 \right) \right] - \frac{\partial \phi}{\partial x^1} K = 0$$

for

$$K = \frac{\partial(x^1, x^2, x^3)}{\partial(x^1_0, x^2_0, x^3_0)}, \qquad K^1_1 = \frac{\partial(x^2, x^3)}{\partial(x^2_0, x^3_0)}, \qquad K^1_2 = \frac{\partial(x^2, x^3)}{\partial(x^3_0, x^1_0)},$$

etc. There is one case in which this expression simplifies. If $\phi(x^1, x^2, x^3)$ can be expressed $\psi(x^1, x^2, x^3)\rho$ where ψ depends only on x^1, x^2, x^3, i.e., the current position, and ρ is the density, then changing variables yields $\psi \rho K$ as the integrand relative to dV_0. But $\rho K = \rho(x_0)$, so that $V = \iiint \psi \rho_0 \, dV_0$, and in the above one can omit the K^1_α terms and replace $(\partial \phi / \partial x^1) K$ by $\rho(\partial \psi / \partial x^1)$.

Presumably the $u_{e1\alpha}$ are functions of the first partials of the x^i relative to the x_0^j so that the first line contains second derivatives relative to both time and the spatial variables. The initial situation for the motion has to be specified.

Clearly the general case is complex. One simplification is associated with assuming that x can be expressed as $x_0 + u$ with u small. The objective is to obtain a simple approximate expression for B. One has

$$J = \left(\delta_j^i + \frac{\partial u^i}{\partial x_0^j}\right), \qquad J' = \left(\delta_j^i + \frac{\partial u^j}{\partial x_0^i}\right),$$

$$B = (JJ')^{1/2} \simeq I + \frac{1}{2}\left(\frac{\partial u^i}{\partial x_0^j} + \frac{\partial u^j}{\partial x_0^i}\right).$$

The components of the second matrix are usually written

$$e_{ij} = \frac{1}{2}\left(\frac{\partial u^i}{\partial x_0^j} + \frac{\partial u^j}{\partial x_0^i}\right).$$

If we consider u_e to be a function of $\mu_i' = \mu_i - 1$, then we can consider u_e to depend on the three functions $p_1 = e_{11} + e_{22} + e_{33}$, $p_2 = e_{11}e_{22} + e_{22}e_{33} + e_{33}e_{11} - e_{12}^2 - e_{23}^2 - e_{31}^2$, and $p_3 = \det(e_{ij})$, since the μ_i' are roots of $x^3 - p_1 x^2 + p_2 x - p_3 = 0$. Since $p_1^2 - 2p_2$ is always positive, u_e is frequently taken in the form $Ap_1^2 + B(p_1^2 - 2p_2)$.

A simplification of practical interest is associated with the experiment used to establish Young's modulus of elasticity. If we take the cube and subject the x^2, x^3 face to a tension T, then $\mu_1 = 1 + T/E$, $\mu_2 = 1 - \sigma T/E$, $\mu_3 = 1 - \sigma T/E$, where T is force per unit area and E is the modulus of elasticity. If $\delta\mu_i = \mu_i - 1$, these equations can be written $E\,\delta\mu_1 = T$, $E\,\delta\mu_2 = -\sigma T$, $E\,\delta\mu_3 = -\sigma T$. If we apply tensions to the other four faces and suppose the effect is additive, we have

$$E\,\delta\mu_1 = T_1 - \sigma T_2 - \sigma T_3$$
$$E\,\delta\mu_2 = -\delta T_1 + T_2 - \sigma T_3$$
$$E\,\delta\mu_3 = -\sigma T_1 - \sigma T_2 + T_3.$$

The equations can be inverted to yield

$$(1 - 2\sigma)(\sigma + 1)T_1 = (1 - \sigma)E\,\delta\mu_1 + \sigma E\,\delta\mu_2 + \sigma E\,\delta\mu_3, \text{ etc.}$$

For metals there is a range of T_i/E for which these relations are of practical significance. Usually this range is limited by, say, $T_i/E < 10^{-3}$. The ratio σ is called Poisson's ratio. One assumes, of course, that the face forces are pure tensions.

7.4. An Elastic Collision

We have as a matter of policy refrained from discussing the solution of partial differential equations, since that is precisely what is readily available in mathematics courses and many other courses. However, it does seem desirable to carry through one example in detail. The example is simplified in the sense that it is based on a one-dimensional version of the three-dimensional discussion given above. This will permit us to handle it by purely formal methods, while in most practical cases, which involve three dimensions, numerical procedures are required.

We consider two steel cylinders or bars with circular cross sections and each of length l. It is convenient to think of l as large relative to the diameter of the cross section. These bars can move lengthwise in a horizontal trough of semicircular cross section in an essentially frictionless manner. The use of an air cushion effect would yield this possibility.

We suppose that initially one bar is standing still and the other bar is approaching it with speed v. The bars collide elastically, that is, without loss of energy. After this, the first bar moves with the original speed v and the second bar is motionless. This result can be readily predicted by Newtonian relativity, but our interest is in what happens in the bars themselves.

We choose the x axis along the line determined by the axes of the cylinder. The origin of this axis is chosen so that the first bar initially is in the interval $0 \leqslant x \leqslant l$. Time is determined so that $t = 0$ corresponds to the instant of first contact, and hence, at $t = 0$ the second bar is located in the interval $-l \leqslant x \leqslant 0$. At this instant the first bar is still, the second bar has speed v directed positively along the x axis.

The collision obviously involves a transformation from kinetic energy into elastic potential energy and then back again into kinetic. The relation governing this transformation is determined experimentally, but it is simpler than that of our previous discussion, since we are interested only in one-dimensional effects. If we take a bar of length h and compress it with a force F to a length $h - e$, then for a steel bar it is found that for a considerable range of F, $F = EA(e/h)$, where E is a constant and A is the area of the cross section. The energy involved in compressing the bar to this extent is $\int_0^e F\, de = EAe^2/2h = Fe/2$. E is very large, so that e/h is usually small. For steel, E is about 2×10^{12} in the cgs system. Thus, if a kilogram of material is supported by a rod of 1 cm^2 in cross section, the weight of the material corresponds to $F = 0.98 \times 10^6$ dynes, so that e/h is about 0.5×10^{-6}.

We take as our reference situation the position of the bars at the instant of contact. For $t \geqslant 0$, the cross section of either bar with abscissa x_0 in the reference position has abscissa $x = x_0 + u(x_0, t)$. We take a subdivision of the initial interval $-l \leqslant x_0 \leqslant l$ with endpoints x_{j-1}, x_j such that the compression in the interval x_{j-1}, x_j is essentially uniform. The uncompressed length of the bar in this interval is $h = x_j - x_{j-1} = \Delta x_j$. The new length is

$$h - e = x_j + u(x_j, t) - x_{j-1} - u(x_{j-1}, t)$$

$$= \Delta x_j + \frac{\partial u}{\partial x} (x'_j, t) \Delta x_j$$

$$= \left(1 + \frac{\partial u'}{\partial x} \right) \Delta x_j$$

so that

$$e = -\frac{\partial u'}{\partial x} \Delta x^j \quad \text{and} \quad F = EA(e/h) = -EA \frac{\partial u}{\partial x} (x'_j, t).$$

Thus, the force across a cross section with initial abscissa x is

$$-EA \frac{\partial u}{\partial x} (x, t) = F(x).$$

For a segment $x_1 \leqslant x_0 \leqslant x_2$ the rate of change of momentum in the positive x_0 direction is

$$\frac{d}{dt} \int_{x_1}^{x_2} A\rho_0 \frac{\partial u}{\partial t} (y, t) \, dy = -F(x_2) + F_0(x_1)$$

$$= EA \left[\frac{\partial u}{\partial x} (x_2, t) - \frac{\partial u}{\partial x} (x_1, t) \right];$$

i.e., the compression on the x_2 cross section will slow the segment down while that on the x_1 cross section will accelerate it. Notice that the right-hand side is really a surface integral. Numerical procedures usually are based on equations of this type.

If we let $c^2 = E/\rho_0$, we have

$$\frac{d}{dt} \int_{x_1}^{x_2} \frac{\partial u}{\partial t} (y, t) \, dy = c^2 \left[\frac{\partial u}{\partial x} (x_2, t) - \frac{\partial u}{\partial x} (x_1, t) \right]. \tag{1}$$

Now if we can find a time interval $t_1 \leqslant t \leqslant t_2$ and a spatial interval $x_1 \leqslant x \leqslant x_2$ such that $\partial u/\partial t$ is continuously differentiable on this interval, this relation yields

$$\int_{x_1}^{x_2} \frac{\partial^2 u}{\partial t^2} (y, t) \, dy = c^2 \left[\frac{\partial u}{\partial x} (x_2, t) - \frac{\partial u}{\partial x} (x_1, t) \right].$$

Thus, for a range of x_2, this equation can be differentiated and yields

$$\frac{\partial^2 u}{\partial t^2} = c^2 \frac{\partial^2 u}{\partial x^2}. \tag{2}$$

The solution of this partial differential equation is in the form

$$u(x, t) = f(x + ct) + g(x - ct),$$

where f and g are arbitrary twice-differentiable functions of a single variable. Unfortunately this is too restrictive. For example,

$$\frac{\partial u}{\partial t}(x, t) = c[f'(x + ct) - g'(x - ct)]$$

and our initial conditions require

$$\frac{\partial u}{\partial t}(x, 0) = v \qquad \text{for } -l \leqslant x < 0,$$

$$\frac{\partial u}{\partial t}(x, 0) = 0 \qquad \text{for } 0 \leqslant x \leqslant l.$$

Hence, $\partial u/\partial t$ is discontinuous, which implies that either $f'(x)$ or $g'(x)$ is discontinuous or both and $\partial^2 u/\partial t^2$ is not available. The solution is to consider Equation (1). We have

$$\frac{\partial u}{\partial x}(x, t) = f'(x + ct) + g'(x - ct),$$

and if f and g are integrals of their derivatives, then the indicated operations on the left-hand side of Equation (1) can be carried out and thus u can be taken in the indicated form with only the restriction that f and g be differentiable and equal to an integral of their derivatives.

Now f and g must be determined by the boundary conditions. These are

(a) $\qquad u(x, 0) = 0 \qquad \text{for } -l \leqslant x \leqslant l,$

(b_1) $\qquad \dfrac{\partial u}{\partial t}(x, 0) = v \qquad \text{for } -l \leqslant x < 0,$

(b_2) $\qquad \dfrac{\partial u}{\partial t}(x, 0) = 0 \qquad \text{for } 0 \leqslant x \leqslant l,$

(c_1) $\qquad \dfrac{\partial u}{\partial x}(-l, t) = 0 \qquad \text{for } t \geqslant 0,$

(c_2) $\qquad \dfrac{\partial u}{\partial x}(l, t) = 0 \qquad \text{for } t \geqslant 0.$

These conditions can be used to determine f and g over the range of interest. Thus, f is defined

$$f(y) = \begin{cases} vy/2c, & -l \leqslant y \leqslant 0 \\ 0, & 0 \leqslant y \leqslant 2l \\ (v/2c)(y - 2l), & 2l \leqslant y \leqslant 3l, \end{cases}$$

and g is defined

$$g(y)=\begin{cases} 0, & 0\leqslant y\leqslant l \\ -vy/2c, & -2l\leqslant y\leqslant 0 \\ vl/c, & -3l\leqslant y\leqslant -2l. \end{cases}$$

Both f and g are equal to integrals of their derivatives, but the derivatives are discontinuous.

The functions f and g describe the situation throughout the collision. The development is most easily visualized by first graphing f, f', g, and g'. The situation at time t is obtained by shifting the graph of f and f' by ct to the left, and that of g and g' by ct to the right.

For the time $0\leqslant t\leqslant l/c$ there are three zones. For $-l\leqslant x\leqslant -ct$ the x cross section is moving with velocity v and has zero stress, and this segment has been displaced to the right an amount vt. For $-ct\leqslant x\leqslant ct$ the x cross section is moving with speed $v/2$, the stress of compression is $Ev/2c$, and the displacement varies uniformly from vt to zero. For $ct\leqslant x\leqslant l$ the displacement, speed, and stress are zero. During this time the second zone expands evenly until at $t=l/c$ the other two zones have length zero.

During the time $l/c\leqslant t\leqslant 2l/c$ there are also three zones. For $-l\leqslant x\leqslant -l+(ct-l)=ct-2l$ the displacement is vl/c, but the speed and stress are zero. For $ct-2l\leqslant x\leqslant 2l-ct$ the speed is $v/2$, the compression is $Ev/2c$, and the displacement varies from vl/c to $vt-vl/c$. For $2l-ct\leqslant x\leqslant 2l$ the speed is v, the compression is zero, and the displacement is $vt-vl/c$. During this time, the middle zone contracts until at time $t=2l/c$ each bar has had a displacement of vl/c, but the lefthand bar has speed zero, the righthand bar has speed v. There is no compression, and the bars begin to separate.

7.5. Thermodynamic States and Reversibility

As discussed in Section 7.2, in the Eulerian description of the behavior of substance, the equation of continuity and Newton's third law yield four equations to determine the quantities ρ, v^1, v^2, v^3. The situation is similar in the alternate approach, i.e., we have three dynamic equations for the $y^i(x_0, t)$, and the density is given in terms of the reference situation density and the Jacobian. But in both cases we must know the a_{ij}, which have to be determined on some empirical basis that must be added to the geometric and Newtonian principles.

The simple experiment we described of compressing the cube of substance and measuring the stresses relative to the given deformation has difficulties that are rather obvious based on general experience. We know that if we compress something, we probably won't receive back the work we did in compressing it. We ignored temperature and the possibility that heat will flow and that this will affect the measured quantities. Actually the situation is quite complex. We must obtain a reliable pattern of experience that can be formulated in mathematical terms. This formulation will involve additional intensive and extensive functions but will be in the same analytic framework of functions, partial differentiation, and geometric integration. This formulation is part of the more general theory of thermodynamics.

We now specialize the a_{ij}. Simplication is desirable and we also need to make contact with the readily accessible literature. We assume that the a_{ij} are in the form $p(y)\delta_{ij}$ and the matrix $a = p(y)I$. For each n, the stress $F(n)$ is along n and at a given point, $F(n)$ has the same magnitude for every direction. This situation is certainly valid in a stationary fluid.

Let $\mathfrak{A}$ be a region in the substance with boundary $\mathfrak{B}$. The force on the substance in $\mathfrak{A}$ due to pressure on the boundary is

$$F_{\mathfrak{B}} = -\iint_{\mathfrak{B}} F(n) dA = -\iint_{\mathfrak{B}} pn \, dA$$

$$= -\iint_{\mathfrak{B}} p(i_1 \, dx^2 \, dx^3 + i_2 \, dx^3 \, dx^1 + i_3 \, dx^1 \, dx^2).$$

If we assume that the body force density is expressible as $\Phi(y)\rho(y)$, then in the Euler description the integral form for Newton's third law is

$$F_{\mathfrak{B}} + \iiint_{\mathfrak{A}} \Phi\rho \, dV = \iiint_{\mathfrak{A}} \rho \frac{DV}{Dt} \, dV.$$

If p has continuous derivatives, this yields

$$-\frac{1}{\rho} \operatorname{grad} p + \Phi = \frac{Dv}{Dt} \, .$$

The component relations of this vector equation are called the Navier–Stokes equations. When there are surfaces of discontinuity for the pressure, the integral form is required. Such surfaces of discontinuity are called shock waves. The situation for the alternate description is analogous.

We now consider the compression experiment with the stress appearing in the form of pressure. We have the associated ideas of heat and temperature. Temperature is an intensive function. Heat is a flow of energy across a surface and occurs when there is a temperature gradient from a hotter to a cooler region. The rate of heat flow may be large or very small in response to a given temperature gradient, depending on the surface. We will suppose that the only actions that occur are compression or expansion and heat exchanges.

This would be the case for an ideal gas and represents the simplest situation in which thermodynamic principles can be formulated. Usually chemical and electrical aspects of energy also must be considered.

Thermodynamics describes the way in which a substance is affected by compression, expansion, and heat exchanges and how it develops spontaneously in various situations. But the basis of this discussion is certain idealized experiments for which a number of thermodynamic functions can be defined and in which these functions have certain mathematical relations. One hopes these relations hold for bodies of very small spatial extent so that they can be used in infinitesimal analysis, and this is the real significance of the "idealization." For experiments involving a finite amount of substance the idealized procedures have two characteristics. One of these is that a body of substance in the experiment must be spatially uniform. Thus, in such a body the density, pressure, and any other intensive function has to be uniform, and any extensive function must have a uniform density. Hence, the development of the experiment can be described by giving the functions as functions of the time, i.e., a curve or path in a space whose coordinates correspond to these functions. A situation in which the intensive functions are uniform in a body is called a thermodynamic state. Thus, each point on the path associated with such an experiment is a state. The second characteristic of the idealized experiments is that if $\mathfrak{C}$ is the path of the experiment in the function coordinate space going from P_1 to P_2, there is also an experiment reversing the path $\mathfrak{C}$ and going from P_2 to P_1. The procedure is then said to be reversible. Irreversible processes include breakage, cold working, or plastic set. One may be able to return to P_1, but not by reversing the path.

For a substance of finite extent the type of change postulated in which the intensive functions vary uniformly in regard to space can be at best only approximated. It is usually assumed that the substance is confined in a cylinder by a piston and that the walls can be modified to permit heat flows. But any motion by the piston or any heat exchange will certainly introduce spatial variations in pressure and temperature. It is assumed that uniformity can be approximated by proceeding very slowly. But this is not the way the relations described in these experiments are verified. Instead the consequences of these relations, especially in infinitesimal analysis, are compared with experience. Clearly these experiments are theoretical concepts utilized in the mathematical formulation of experience.

7.6. Thermodynamic Functions

Consider, then, such a piston–cylinder–substance apparatus that varies only in such a way that the substance is always in a thermodynamic state.

For the substance there must be one extensive quantity, and there is an advantage in taking it to be the mass, which is a constant. The intensive quantities clearly include pressure, density, and temperature. For reasons that will be apparent it is customary to replace the density by the volume of substance, i.e., the mass divided by density. There are circumstances in which this replacement is awkward. However, two other intensive functions appear, the internal energy per unit mass and the entropy per unit mass. To agree with the usual notation, which implies that one is dealing with the corresponding extensive functions, we will take the mass as a unit and let U stand for the internal energy and S for the entropy.

There are two ways in which the substance participates in energy exchanges. One of these is the work done by the substance by moving the piston, say, an amount ds. Then $dW = F\, ds = pA\, ds = p\, dV$. We suppose that we can control the nature of the walls of the cylinder relative to the transmission of heat. The amount of heat entering the substance is denoted by dQ. The principle of conservation of energy states that $dU = dQ - p\, dV$. Since there are two independent ways in which U can be varied, the possible paths for a thermodynamic process lie in a two-dimensional surface. In many cases this surface can be parametrized by choosing two of the intensive functions as parameters and expressing the others in terms of them, and this is what is usually done. One such choice of a pair of functions is the pressure and volume, p, V.

Consequently the differentials we deal with are differential forms on two independent variables. The differentials dW and dQ are such forms, but dU is an exact differential. We also have in the coordinate function space for the substance an exact differential dS such that $dQ = T\, dS$, where T is the temperature and S is the entropy.

An interesting example is represented by an "ideal gas." There are certain relations between the functions for this theoretical substance, and they are derived from a kinetic model of a gas. For an actual gas these relations are verified experimentally, and they usually apply if the gas is not near condensation. These relations are $pV = MRT$ and $U = \frac{3}{2}MRT$, where T is the absolute temperature and R is a constant chosen so RT has a value in the appropriate energy units. Notice that these relations can be written, $p = \rho RT$ and $u = \frac{3}{2}RT$, where ρ is the density and u is the energy per unit mass. We will follow the more usual choice of function, that is, V, and also will take $M = 1$.

We have then

$$T\, dS = dQ = dU + p\, dV = \tfrac{3}{2} R\, dT + p\, dV$$

or

$$dS = \frac{3}{2} R \left(\frac{dT}{T} + \frac{2}{3} \frac{p\, dV}{RT} \right) = \frac{3}{2} R \left(\frac{dT}{T} + \frac{2}{3} \frac{dV}{V} \right)$$

or

$$S = \tfrac{3}{2} R \ln (TV^{2/3}) + C = \tfrac{3}{2} R \ln (pV^{5/3}) + C',$$

where C and C' are constants.

Of course, the assumption that we have used earlier that there is an elastic energy function means that the coordinate function space is one-dimensional. For this type of situation we must have $dQ = 0$, that is, $dS = 0$ or S is a constant. The system then must vary in the p, V plane along a curve $pV^{5/3} = A$. If the substance varies from a point P_1 with coordinates p_1, V_1 and temperature T_1 to a point $P_2, p_2, V_2,$ and T_2, then the work done by the substance is

$$W = \int p \, dV = - \int dU = U\big|_2^1 = \tfrac{3}{2} R(T_1 - T_2).$$

The alternate form of the adiabatic condition, $TV^{2/3} = B$, yields

$$\left(\frac{V_1}{V_2}\right)^{2/3} = \frac{T_2}{T_1}.$$

The adiabatic assumption is used in treating the propagation of sound, since the oscillations are fast enough so that the immediate heat flow can be neglected.

In general the "substance" may have a more complex structure, and dU may have additional terms. For example, a voltage cell may produce a transfer of electric charge and one adds a term $v(-de)$, where v is the voltage. A major practical aspect of energy transformation is chemical, and one has composition ratios η_i and "chemical potentials" μ_i that constitute terms $\mu_i \, d\eta_i$, which are added to dU. Electric and magnetic fields may induce polarization, magnetization, and electric currents. Thus, the dimensionality of the intensive function coordinate space may be much higher. All these energy forms act in conjunction with heat flows. Heat flows can be between bodies in contact or by radiation. In general changes in U will correspond to gradients in the coefficients p, T, v, μ_i, etc. or differences in their values for different bodies. Rates of change usually require additional empirical information.

7.7. The Carnot Cycle and Entropy

We consider again the case of a single substance whose two-dimensional function coordinate space can be parametricized in terms of pressure and volume, p and V. Suppose we have such a substance in a cylinder with a piston and two reservoirs for heat, one at a higher temperature than the other.

We suppose that we have controlled heat-conducting connections between the reservoirs and the substance in the cylinder. There is a procedure, A, by which heat can be transferred through the cylinder–piston–substance apparatus from the hot reservoir to the cool one and partly converted into work W_A. There is also a procedure B by which an amount of work W_B can be used to produce a heat interchange in the opposite direction. In practice, however, if we apply A and B or B and A successively so as to restore the original situation in the reservoirs and apparatus, then $W_B > W_A$. There are always uncontrolled heat flows, which account for this difference as an increase in entropy.

There is, however, a thermodynamically idealized version of procedure A that is "reversible." This is termed the Carnot cycle. We will suppose that the apparatus substance is a perfect gas, and the time development of this gas will be described by curves in the plane with Cartesian coordinates p and V. This cycle has four phases, each of which corresponds to a segment of a curve. We begin the first phase at a point P_1, p_1, v_1, which is at temperature t_1, the same as the cooler reservoir. In the first phase the substance is compressed adiabatically from a volume v_1 to a volume v_2, the abscissa of the point P_2, p_2, v_2 so that the temperature increases from t_1 to t_2, the temperature of the hot reservoir, in accordance with formula

$$\left(\frac{t_2}{t_1}\right) = \left(\frac{v_1}{v_2}\right)^{2/3}$$

from the previous section. The work done by the substance in this phase is negative, $W_1 = \frac{3}{2}R(t_1 - t_2)$. By means of an infinitesimal gradient heat is now permitted to flow from the hot reservoir into the substance that remains at the same temperature t_2. In this second phase, U for the gas is a constant, and thus the work done, W_2, by the substance equals the heat, Q_2, received, and furthermore, for $M = 1$ we have

$$W_2 = \int_{v_2}^{v_3} p\, dV = Rt_2 \int_{v_2}^{v_3} \frac{dV}{V} = Rt_2 \ln\left(\frac{v_3}{v_2}\right).$$

The third phase is an adiabatic expansion from v_3 ro v_4 corresponding to a return of the gas to temperature t_1; i.e., we have $t_2/t_1 = (v_4/v_3)^{2/3}$ and $W_3 = \frac{3}{2}R(t_2 - t_1)$. We again have a isothermal contraction in the fourth phase with heat Q_4 flowing from the substance into the cool reservoir, $Q_4 = -W_4$, and

$$W_4 = -Rt_1 \ln(v_4/v_1).$$

This isothermal contraction restores the original volume v_1 at temperature t_1, and hence, the original pressure p_1 is also restored. This cycle is illustrated in Figure 7.1.

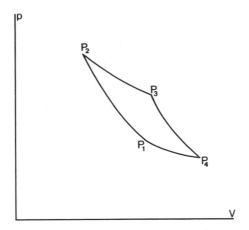

Figure 7.1. The Carnot cycle.

Thus, in this process the apparatus is restored to its original condition, an amount $Q_2 = Rt_2 \ln(v_3/v_2)$ of heat has come from the hot reservoir, and amount $Q_4 = Rt_1 \ln(v_4/v_1)$ has entered the cool reservoir, and the difference $Q_2 - Q_4$ corresponds to the total work done by the substance. Since $v_2/v_1 = (t_1/t_2)^{3/2} = v_3/v_4$, we have $v_3/v_2 = v_4/v_1$; the output work is

$$W = W_2 + W_4 = Q_2 - Q_4 = R(t_2 - t_1)\ln(v_3/v_2).$$

The hot reservoir has had a decrease in entropy $Q_2/t_2 = R \ln(v_3/v_2)$ and the cool reservoir an increase in entropy Q_1/t_1, which has this same value. This whole procedure can be reversed by cycling in the inverse order. Thus, one could continue to convert heat into work until the temperatures are equal or, conversely, produce a heat difference by work.

It is interesting to see what one should expect in practice if one tried to realize this set up. Suppose that in regard to heat the combination of apparatus and two reservoirs are completely isolated. There would be, of course, heat leaks from the hot reservoir to the cool one, and if an amount of heat Q leaked in this way, the two reservoirs would have an increase in entropy $Q(1/t_1 - 1/t_2) = Q(t_2 - t_1)/t_1 t_2$. In the adiabatic compression stage the compressive work would exceed the actual increase in energy U, and presumably the extra work would be converted to heat leaking into the cooler reservoir. In the heat absorption stage the actual temperature of the substance is less than that of the reservoir to induce a flow of heat, and hence, the entropy gain of the substance is greater than the entropy loss of the upper reservoir. There is a converse result in the fourth stage so that the increase of entropy in the cool reservoir exceeds the loss of entropy of the substance. In the adiabatic expansion the work obtained will be less than the change in U. Thus, in general one can consider the actual operation as that of the theoretical Carnot cycle with a leak, and this permits a definition of efficiency as the ratio of output work to the difference $Q_2 - Q_4$. When Q_2 has been obtained by the consumption of fuel and the lower reservoir is considered to be essentially infinite in extent, it is more customary to define efficiency as the ratio output work to Q_2.

7.8. The Relation with Applied Mathematics

The procedures of the calculus based on such notions as functions, partial differentiation, and geometric integration constitute a mathematical format that is applicable to three major aspects of the exact sciences. These are Newtonian dynamics (including the variational developments), electromagnetism, and thermodynamics.

A culture has many aspects, but there must always be a mainstream of activity that provides the necessities and amenities of life, and there must also be communication. In our culture these three scientific areas structure the technology for this mainstream and communication. This is evident in the role of machinery for production and transportation and in our methods for communication and handling power.

Scientific understanding has a unity that does not permit facile dissection into subareas. The word "aspect" is perhaps most appropriate to indicate the role of thermodynamics in applications. Except for water power, our major methods of producing energy depend on first converting a latent source of energy into heat, second, converting this heat into mechanical energy, and third, converting the mechanical energy into electrical. The understanding of the second step in this process was both of great practical value and philosophical significance. This understanding was required both for the efficient production of power and for the use of power in transportation.

The universality of the first and second laws of thermodynamics colors the thought of all persons who have grasped its meaning. If like the ancient philosophers one seeks for a single element that is the fundamental constituent of everything, then there is only one candidate available in light of our present knowledge—that is energy. The laws of thermodynamics are the rules that describe the course of energy even to the most esoteric extremes, far beyond the usual experience that we have been discussing.

A very important application of thermodynamics is to structure the energy relations of chemistry. Thermodynamic considerations limit sharply the amount of empirical information needed to establish the energy of composition of chemical compounds. These considerations also yield equilibrium relations and reaction rates. Physical chemistry and thermodynamics are essentially inseparable.

We see, then, a continued complementary development of cultural capability and the mathematical formulation of experience. The elementary combination of finite set logic and procedures with natural numbers form the basis for mercantile transactions and the economic aspects of social relations. Geometry represented a highly significant expansion of this capability to include magnitudes such as lengths, areas, and volumes. The problem of incommensurability appears at first to be simply a philosophical question and not one of practical significance. But a satisfactory logical structure for the expanded mathematical development proved to be of utmost importance. We have seen how calculus (that is, infinitesimal analysis) greatly expanded the area of experience subject to mathematical formulation.

We have emphasized comprehension in mathematical terms rather than the subsequent procedures for solving the differential equations that result from infinitesimal analysis. But it was the availability of such methods that focussed the approach we have discussed. In recent years the limitations of classical analysis were considerably relaxed by the development first of analog computing and then by digital numerical methods.

The actual temporal development of these areas in the exact sciences was quite complex and subject to many cross currents. The notion of what constituted a satisfactory logical formulation of mathematics gradually emerged. But this was also associated with a very flexible system of new mathematical concepts that were suitable for handling an expanding range of experience.

The classical "method of fluxions" is confined to gross phenomena in which the fine structure of matter has only an averaged effect. The effort to understand this fine structure in classical terms ran into very serious difficulties, and new mathematical concepts were required. There were a number of motivations for trying to understand the fine structure. One of these was precisely the inconsistencies resulting from the classical approach. Another was the desirability of replacing the empirical relations needed to supplement classical methods by a theoretically consistent development. There were also technological rewards, such as the possibility for new and extraordinary weapons and communication devices.

For further discussion of the material in this chapter, see Bergmann,[1] Denbigh,[2] Fermi,[3] Lamb,[4] Love,[5] Mahan,[6] and Murnaghan.[7]

Exercises

7.1. An airplane makes a tight turn so as to "pull 8 g's." What was the minimum radius of curvature?

7.2. What does the expression "sidereal month" mean? Based on this information, how far is the moon?

7.3. An automobile traveling on a level road is brought to a quick stop by applying the brakes. Discuss quantitatively the forces, torques, and angular motions involved.

7.4. For $J = (\delta^i_j + \partial u^i/\partial x^j)$, $J' = (\delta^i_j + \partial u^j/\partial x^i)$, one has $B = (JJ')^{1/2}$. Let r, s, t be the elementary symmetric functions of JJ'; i.e., if α_1, α_2, α_3 are the characteristic roots of JJ' one has $r = \alpha_1 + \alpha_2 + \alpha_3$, $s = \alpha_1\alpha_2 + \alpha_2\alpha_3 + \alpha_3\alpha_1$, and $t = \alpha_1\alpha_2\alpha_3$. Let p_1, p_2, p_3 be the corresponding quantities for B. One can show that $p_1^2 = r + 2p_2$, $p_2^2 = s + 2p_1p_3$, and $p_3^2 = t$ and that there is a quartic equation for p_1, with coefficients expressed in terms of r, s, and t.

7.5. Suppose one has an elastic energy function $W(p_1, p_2, p_3)$ expressed in terms of the elementary symmetric functions of the characteristic roots μ_1, μ_2, μ_3 of B. Then a has the same characteristic vectors and characteristic roots $\lambda_i = (\partial W/\partial \mu_i)\mu_i/p_3$. Then

$$a = \frac{1}{p_3}\left[\left(\frac{\partial W}{\partial p_1} + p_1\frac{\partial W}{\partial p_2}\right)B - \frac{\partial W}{\partial p_2}B^2\right] + \frac{\partial W}{\partial p_3}I.$$

7.6. Let T be an $n \times n$ matrix of real numbers. Show that there is an orthogonal matrix O and a positive definite symmetric matrix A such that $T = OA$. Are O and A uniquely determined by T?

7.7. If A is a symmetric matrix of real numbers, there is an orthogonal matrix Q such that QAQ^{-1} is diagonal. Prove this and discuss the extent to which Q is determined.

7.8. Given

$$T = \begin{pmatrix} 1.07 & 0.1 & 0.05 \\ -0.05 & 0.8 & -0.03 \\ 0.07 & -0.15 & 0.75 \end{pmatrix},$$

find A, B, and O for T. (A and B are positive definite, O is orthogonal, and $T = OA = BO$.)

7.9. For J in Exercise 4, write $J = I + j$, $J' = I + j'$. Then $B^2 = 1 + j + j' + jj'$. It would be desirable to express an elastic energy function in terms of the invariants of $j + j'$ that is the coefficients of its characteristic equation. But B^2 is not determined by $j + j'$, so this is not justified. [*Hint*: Consider the case in which $j' = -j$.]

7.10. A metal ring whose cross section has a diameter small relative to the radius of the ring is rotating around its center. Show that the stress on a cross section is $v^2\rho$, where ρ is the density and v the linear velocity of a point on the ring. How much will it expand if it is steel? aluminum? What are the highest speeds that can be obtained?

7.11. A metal disk of uniform thickness rotates around its center with angular velocity ω. Let the reference position be that in which $\omega = 0$. Then the point with cylindrical coordinates r, θ, z in the reference position will move to the point $r + u(r)$, θ, z when the disk is rotating. A small piece of the disk will be extended in both the radial direction and in the local direction of motion. Suppose that this is due to two simple tensions in these directions. Then u satisfies a differential equation

$$\left(\frac{u}{r} - \sigma\frac{du}{dr}\right)\left(1 + \frac{du}{dr}\right) - \frac{d}{dr}\left[(r+u)\left(\frac{du}{dr} - \sigma\frac{u}{r}\right)\right] = C(r+u)u,$$

with $C = \rho_0 E\omega^2/(1-\sigma^2)$. This can in general be linearized by neglecting second-degree terms in u and its derivatives.

7.12. If $J = (\partial y^i/\partial x^j)$ is the matrix for a change of variables, $y^i = y^i(x^1, x^2, x^3)$, then $J = jA$, where $j = (j^i_j)$, is orthogonal and A is positive definite. The orthogonal matrix j

is a function of the x position and constitutes a field of orthogonal matrices. Let

$$\omega_{ijk}=j^\alpha_i\frac{\partial j^\alpha_j}{\partial x^k}, \quad \omega_k=(\omega_{ijk}).$$

Then $dj=j\omega_\alpha\,dx^\alpha$, ω_k is antisymmetric, and

$$\frac{\partial\omega_{ijk}}{\partial x^l}-\frac{\partial\omega_{ijl}}{\partial x^k}+\omega_{i\alpha l}\omega_{\alpha jk}-\omega_{i\alpha k}\omega_{\alpha jl}=0$$

or, in terms of matrices,

$$\frac{\partial\omega_k}{\partial x^l}-\frac{\partial\omega_l}{\partial x^k}+\omega_l\omega_k-\omega_k\omega_l=0.$$

If a field of ω_{ijk} satisfies this system of partial differential equations, it determines a field of orthogonal matrices up to a multiplicative constant matrix. There are similar results for the $\omega^i_{jk}=j^i_a\,(\partial j^i_\alpha/\partial x^k)$. [*Hint*: One has the system of partial differential equations

$$j^i_\alpha\omega_{\alpha jk}=\frac{\partial j^i_j}{\partial x^k}$$

for the j^i_j values in terms of the ω_{ijk}.]

7.13. In Exercise 12 one has the relations $\partial y^i/\partial x^j=j^i_\alpha a_{\alpha j}$ for $A=(a_{ij})$. These imply

$$\omega_{i\alpha k}a_{\alpha j}-\omega_{i\alpha j}a_{\alpha k}+\frac{\partial a^i_j}{\partial x^k}-\frac{\partial a^i_k}{\partial x^j}=0.$$

Let a symbol $[i,j]$ be defined by $[1,2]=3$, $[2,3]=1$, $[3,1]=2$ and let curl a denote the matrix with components

$$(\text{curl }a)_{i[j,k]}=\frac{\partial a^i_j}{\partial x^k}-\frac{\partial a^i_k}{\partial x^j}.$$

Let a_i denote the vector corresponding to the ith row of the matrix A and $(\text{curl }a)_i$ be similar. Let $\sigma^{[i,j]}$ denote the vector $(\omega_{ij,1},\,\omega_{ij,2},\,\omega_{ij,3})$ and $a^{[ij]}=a_i\times a_j$. Then the above ω, a relation is equivalent to

$$a^i\cdot\sigma^j-\delta^i_j(\sigma^\alpha\cdot a^\alpha)+(\text{curl }a)_i\cdot a_j=0$$

or

$$a^i\cdot\sigma^j-\delta^i_j[\tfrac{1}{2}(\text{curl }a)_\alpha\cdot a_\alpha]+(\text{curl }a)_i\cdot a_j=0.$$

Since $a^i\cdot a_j=\delta^i_j\det(A)$, this yields

$$\det(A)\sigma^j=[\tfrac{1}{2}(\text{curl }a)_\beta\cdot a_\beta]a_j-[(\text{curl }a)_\alpha\cdot a_j]a_\alpha.$$

Thus, the ω_{ijk} are determined by the matrix A. The condition of the previous exercise now becomes a condition on the matrix A, and the matrix A determines the j field up to a multiplicative constant.

7.14. By considering the inverse change of variables one can show that the B matrix for which $J=Bj$ also determines the field of orthogonal matrices j^{-1} up to a multiplicative constant matrix and hence j. Here j is a field on the y space.

7.15. A cylinder is said to be uniformly twisted by the transformation $z'=\alpha z$, $r'=\beta r$, $\theta'=\theta+\gamma z$. Find the matrix B, its characteristic vectors and values, and the corresponding stresses using the E, σ formulas.

7.16. Consider an acoustic plane wave in a perfect gas. We suppose that the gas is in a cylindrical container with elements parallel to the x axis. The acoustic wave consists of displacing the substance in such a way that each cross section with abscissa x in the reference position is displaced to a position with abscissa $x+u(x,t)$. The pressure density relationship is adiabatic, i.e., $p=p_0\rho^{5/3}/\rho_0^{5/3}$. One obtains the relations

$$\rho_0 \frac{\partial^2 u}{\partial t^2} = -\frac{\partial p}{\partial x}$$

$$\rho \left(1+\frac{\partial u}{\partial x}\right) = \rho_0$$

$$\left(1+\frac{\partial u}{\partial x}\right)\rho_0 \frac{\partial^2 u}{\partial t^2} = \frac{5}{3} p \frac{\partial^2 u}{\partial x^2}.$$

7.17. Develop a scenario corresponding to what one would expect if one tried to realize a Carnot cycle. What are the considerations involved in using a p, V diagram? How would efficiency be expected to vary with output power?

7.18. Suppose one has the Carnot apparatus and cycles it through a closed curve in an oval shape in the p, V plane (see Figure 7.2). Let P_1, P_2 be the points where the tangent is parallel to the y axis, Q_1, Q_2 the points where $dp/dV = -5p/3v$, and R_1R_2 the points where $dp/dv = -p/v$. Suppose a cycle in the clockwise direction reaches the points $R_1Q_1P_1, R_2Q_2P_2$ in this order. Then the substance does work in going from P_1 to P_2 and negative work in going from P_2 to P_1. The substance absorbs heat in going from Q_1 to Q_2 and emits heat in going from Q_2 to Q_1. The internal energy will increase in going from R_1 to R_2 and decrease in going from R_2 to R_1.

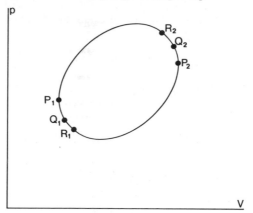

Figure 7.2. Cycling points.

7.19. Discuss the energy exchanges associated with our various sources of power. How does entropy increase and what happens to this increase?

References

1. Bergmann, P. G., *Mechanics and Electrodynamics*, Dover Publications, Inc., New York (1962).
2. Denbigh, Kenneth, *The Principles of Chemical Equilibrium*, Cambridge University Press, Cambridge, England (1964).

3. Fermi, Enrico, *Thermodynamics* (reprint), Dover Publications, Inc., New York (1937).
4. Lamb, Sir Horace, *Hydrodynamics* (reprint), Dover Publications, Inc., New York (1945).
5. Love, A. E. H., *The Mathematical Theory of Elasticity* (reprint), Dover Publications, Inc., New York (1945).
6. Mahan, Bruce H., *Elementary Chemical Thermodynamics*, W. A. Benjamin, Inc., Menlo Park, California (1964).
7. Murnaghan, F. D., *Finite Deformations of an Elastic Solid*, Dover Publications, Inc., New York (1967).

8

Probability

8.1. The Development of Probability

The theory of probability appears to have risen from two sources, both of great antiquity—marine insurance and games of chance. Marine insurance was practiced by the Babylonians, Phoenicians, Rhodians, Greeks, and Romans. It persisted through the Dark Ages and medieval times. The English and other northern Europeans followed Italian models in the sixteenth century, and this business ultimately expanded worldwide. Rates were set initially by individual insurers, but later much more sophisticated procedures were required (see Flower and Jones[10]).

The formal development of the theory of probability began, according to Todhunter,[17] with a certain correspondence, about 1654, between, Pascal, Fermat, and the Chevalier de la Mere concerning games of chance. Treatises were written by Huygens (1657), James Bernoulli (1705), Montmort (1708), and De Moivre (1711). The increasing use of analytic methods during the eighteenth century is represented by works of Daniel Bernoulli (1783), Euler (1764–1766), Bayes (1763–1765), Lagrange (1770–1773), and Laplace (1774). Modern references are Feller[9] and Cramer.[6]

In the probability theory for dice or card games, there is supposed to be for each play, i.e., die throw or hand of cards, a totality of equally likely events and a subset of favorable events, and the ratio of the number of the latter to that of the totality is the probability of a favorable outcome. In playing the game one has a set of compound events with varying possible outcomes, and the mathematical theory is analogous to a measure and integration theory on finite discrete sets with a multiple dimension structure. The technical mathematical problem is to obtain closed formulas for the count of various possibilities. The binomial coefficients, $\binom{n}{k}$, specify the ways in which one can choose k elements from a set of n, and Pascal invented the well-known triangle for computing these inductively. In general, quite ingenious analogies, especially with algebraic processes, are possible. Thus, if a die is thrown n times, the number of sequences of throws that will have p as the sum is the coefficient of x^p in the expansion of $(x + x^2 + x^3 + x^4 + x^5 + x^6)^n$, for a choice of a term from each of the n factors of the last expansion is analogous to a sequence of n throws of the die.

Elementary probability theory is a form of measure theory on a structured set of events. The rules for specifying the measure and integrals in accordance with the structure constitute a mathematical theory that appears to be applicable to various areas of experience. The question is, when is this theory applicable?

In many instances we are entirely confident about such applications. In these cases we seem to have a natural extension of the conceptual process by which events are distinguished into "objects" to which set theory (or formal logic) are applicable. This additional aspect yields that each of a certain set of events are "equally likely." Presumably this is a long-range intuitive integration of experience. The predictions of probability theory are themselves probability statements and have an ambiguity that we have learned to live with. Do I take an umbrella when the weather prediction is for a "20% chance of precipitation"?

But there are also many situations where this intuitive assignment of probability is not available to us, and the mathematical theory of probability has been developed to cope with this. In modern science, probability theory provides the essential concepts for structuring empirical information into a preliminary scaffolding of understanding that may be subject to further development. If two events in our experience are associated only by chance, we cannot influence the second by dealing with the first. If the use of a certain medicine does not correspond to a change of the frequency of recovery from a specified disease, it is worthless for this purpose. But where chance occurs, the probability measure is often of great practical significance, as we see in the case of insurance.

Thus, in the development of understanding, the probability concept is very valuable as guidance for the action required. Consequently the motive for the theory of probability is either to show the existence of nonchance relationships between events or where chance relations do hold to establish the characteristics of the probability measure. Probability theory makes only probability predictions, but if a hypothesis about the probability distribution in certain circumstances predicts that a specified outcome of an experiment has only small probability but the experiment consistently yields this outcome, we may reject the hypothesis. The practical difficulty is the word "consistently," and here judgment with a certain connotation of arbitrariness is required. If the whole situation can be immersed in an intuitive probability framework so that an *a priori* probability for the hypothesis is available, Bayes' rule will yield answers within this framework. But ultimately the nature of probability itself requires that there will always be a residue of arbitrary decision in the acceptance of a theory on an empirical basis.

Hence, in general the mathematical theory of probability is concerned with a sequential or multidimensional structure of events and the appropriate measures and integrals. The objective is to obtain "asymptotic results," i.e., methods for approximating limits by taking n large where n is either the sequence subscript or the dimension. This theory is immediately available in the references, i.e., Todhunter,[17] Feller,[9] or Cramer.[7] The student is probably aware of the meaning of such technical terms as distribution function, frequency, mean, expected value, variance, and likelihood. The asymptotic theory is dependent on the use of Stirling's formula and the Fourier transform. Major distinctions are made between discrete events and continuous sets of events and in the latter between the normal distribution case and the more general case in which normality is not assumed. The theory of statistics is structured by the objectives of determining nonprobabilistic relations and, where probabilistic associations are known, of determining the characteristics of the probability distribution.

Todhunter's history describes a considerable variety of games of chance following the example of Montmort, who believes this to be necessary because *"pour l'ordinaire, les Scavans ne sont pas Joueurs."* But in addition there are applications to lotteries, life insurance, annuities, demography, errors in experiments, the incidence of smallpox and the effect of vaccination on smallpox, which was the dangerous procedure that preceded the relatively safe vaccination with cowpox of Jenner. There was also considerable philosophical speculation using probability.

For further discussion of the material in this section, see Cramer,[6],[7] Feller,[9] Flower,[10] and Todhunter.[17]

8.2. Applications

Of course, the modern range of applications of probabilities is more extensive. In matters of skill probability evaluations are usually given. For example, in baseball one has batting averages, fielding averages, earned run averages, etc. Reliability of manufactured products is expressed as the probability of the product being free of defects. The biological models for hereditary, mutation, and survival are based on probability. The modern theory of statistics developed in association with experimentation in agriculture.

Testing drugs and medical procedures such as vaccination are also examples of applied statistics.

Much of mathematical analysis is formulated in terms of a choice or strategy. In a maximum problem in the calculus one has a situation in which desirability is measured by the value of a function whose variable x can be specified. This can be considered a "one-person" game. In a two-person game the outcome, say, the amount won by one player, will usually depend on choices by both players. This is analogous to considering the outcome as a function of two variables, x and y, which the first player tries to maximize by his choice of x and which the second player tries to minimize by his choice of y. In a smooth case the play would correspond to a saddle point for the outcome function. When more than two players are involved, there is an additional possibility for cooperation by a proper subset of the players.

The theory of such strategies is called the theory of games. It has been extensively studied as a basis for models for economic theory. There is also considerable interests in winning strategies for recreational games (see Blackwell and Gershick,[1] von Neumann and Morganstern,[19] or Burger[4]).

The critical development of statistics occurred early in the twentieth century. This was one element in the tremendous increase in sophistication of experimentation in the natural sciences, physics, biology, and chemistry. Another aspect was the understanding of electromagnetism and light and the use of electromagnetic radiation. Improvements in instrumentation such as the diffraction spectroscope and electronic amplification were balanced by mathematical procedures using probability theory and integral transform analysis. In the 1950s, large-scale automatic electronic computation added new dimensions. The exploration of the physical structure of the cell and the chemical life processes was extremely effective. This utilized the electron microscope and X-ray diffraction analysis. The latter had been applied initially to inorganic crystals but was expanded by the use of Fourier series to yield the structure of remarkably complex organic molecules (see Bragg[3]).

For further discussion of the material in this section, see Blackwell and Gershick,[1] Bragg,[3] Burger,[4] and von Neumann and Morgenstern.[19]

8.3. Probability and Mechanics

At first glance it may seem that a probability description of a toss of a coin and a dynamic description based on Newtonian physics are antithetical. One could apply either but not both. However, the two descriptions can be readily reconciled. When we toss the coin we do not precisely determine the initial conditions of the motion. The outcome of the toss is clearly dependent on the state of motion and position of the coin as it leaves the hand, since the motion after this point is clearly that of a rigid body. There may be some elastic reaction when landing. Presumably there is a probability distribution for these initial conditions, and a more complete mathematical description of the situation would include this probability distribution for the initial conditions as well as the dynamic description of the motion. Thus, the two aspects are complementary parts of a more complete mathematical theory.

This complementary characteristic is valid over a wide range. Suppose a system is governed by a Hamiltonian,

$$H(p, q), \qquad p_i, q^i, \qquad i = 1, \ldots, n,$$

$$\dot{p}_i = \frac{\partial H}{\partial q^i}, \qquad \dot{q}^i = -\frac{\partial H}{\partial p^i}.$$

Let π denote the differential $\Pi_i\, dp_i\, dq^i$, and let π_i denote the result of omitting dp_i and π^i the result of omitting dq^i. Let $\mathfrak{B}$ be a surface that bounds a region $\mathfrak{A}$. Then the rate at which the volume of $\mathfrak{A}$ changes is given by

$$2n\frac{dV}{dt} = \frac{d}{dt}\iint_{\mathfrak{B}} \sum_{i=1}^{n} (p_i\pi_i - q^i\pi^i) = \iint_{\mathfrak{B}} \sum_{i=1}^{n} (\dot{p}_i\pi_i - \dot{q}^i\pi^i)$$

$$= \iint_{\mathfrak{B}} \sum_{i=1}^{n} \left(\frac{\partial H}{\partial q^i}\pi_i + \frac{\partial H}{\partial p_i}\pi^i\right)$$

$$= \iiint_{\mathfrak{A}} \sum_{i=1}^{n} \left(\frac{\partial^2 H}{\partial q^i\partial p_i} - \frac{\partial^2 H}{\partial p_i\partial q^i}\right)\pi$$

$$= 0.$$

Thus, the volume in the phase space $\{p, q\}$ is invariant under the motion.

This volume in phase space is not a probability despite its apparent measure character. It should be regarded as the equivalent to a count of possibilities or of "events" like the faces of a die. Since the total measure is infinite, one cannot take the probability as simply proportional to this measure. In modern developments this Liouville measure is replaced by discrete counts of possibilities, but even here these possibilities cannot be considered equally likely.

The assignment of probabilities to the possibilities associated with a region in phase space at an instant of time is dependent on the availability of energy. Suppose we have a large number, N, of systems of the above type that are in thermal equilibrium with the rest of the universe, i.e., have a net zero exchange of energy with the remainder of the universe. Furthermore, these systems interact with each other only sporadically.

We divide the phase space of an individual system into a number of small regions, R_i, each with approximately constant energy ε_i. We ask what is the probability that the N systems can be distributed among these small regions so that n_i of them are in the region R_i.

To each region R_i we assign a weight ω_i, which we consider to be a count of the number of possible states that are in the region R_i. This weight ω_i is proportional to the volume of phase space V_i for R_i in the case of a classical system. We assume that the volume h^n corresponding to one state is so small that $\omega_i = V_i/h^n$ can be considered a relatively large integer.

We make a number of other classical assumptions. We suppose that we can distinguish the individual systems so that the number of ways in which the N systems can be divided into sets with numbers $n_1, n_2, \ldots$ is given by $N!/n_1!n_2!\ldots$. Each of the n_i systems assigned to the R_i region can be placed in ω_i different states, so that the number of possibilities for the proposed distribution is

$$v(n_1, n_2, \ldots) = N!\omega_1^n\omega_2^n \cdots /n_1!n_2!\cdots.$$

If only one system can be assigned to a given state, $\omega_i^{n_i}$ is to be replaced by $\omega_i!/(\omega_i - n_i)!$, but this will not make much difference if ω_i is large relative to n_i.

But the energy equilibrium requirement states that only distributions $n_1, n_2, \ldots$ with a given total energy $NU = \Sigma_i\, n_i\varepsilon_i$ should be considered and, of course, we must have $N = \Sigma_i\, n_i$. Subject to these restrictions we can regard all such possibilities as equally likely, and $v(n_1, n_2, \ldots)$ is a relative (unnormalized) probability. Notice that suitable restrictions on the ε_i will insure that the total number of possibilities is finite.

Given the values of N and U, the obvious objective is to normalize the relative probability v. However, the practical procedure in the "classical" case of ω_i large is governed by the fact that there is a most probable set of values $n_1, n_2, \ldots$ and that for N large the probability of a distribution with values significantly different from the most probable values is very small. Thus, the practical procedure consists in assuming that the N systems are distributed as they are in the most probably distribution.

To determine the most probably set of values for the $n_1, n_2, \ldots$ one must maximize $v(n_1, n_2, \ldots)$. Consider

$$\log v(n_1, n_2, \ldots) = \log N! - \sum \log n_\alpha! + \sum n_\alpha \log \omega_\alpha.$$

We will simply lump additive constants. By Stirling's formula $n! \eqsim n^n \exp(-n)/(2\pi n)^{1/2}$, and

$$\log v = -\sum_\alpha \left[(n - \tfrac{1}{2}) \log n_\alpha - n_\alpha + n_\alpha \log \omega_\alpha\right] + C.$$

This expression must be maximized subject to the side conditions, $\Sigma\, n_\alpha = N$,

$\Sigma\, n_\alpha \varepsilon_\alpha = NU$. This requires constant α and β such that

$$\frac{\partial \log v}{\partial n_i} = -\alpha + \beta \varepsilon_i$$

or

$$1/2n_i - \log n_i + \log \omega_i = -\alpha + \beta \varepsilon_i.$$

Neglecting the term $1/2n_i$ we obtain, for $A = e^\alpha$,

$$n_i = \omega_i A \exp(-\beta \varepsilon_i).$$

If we let $F(\beta) = \Sigma \omega_\alpha \exp(-\beta \varepsilon_\alpha)$, we obtain by adding these equations $N = AF(\beta)$. Thus,

$$\frac{n_i}{N} = \frac{\omega_i \exp(-\beta \varepsilon_i)}{F(\beta)}.$$

This can be interpreted as a statement that the probability that a system be in the region R_i is $\omega_i \exp(-\beta \varepsilon_i)/F(\beta)$, with β determined by the condition

$$NU = \sum_\alpha \varepsilon_\alpha n_\alpha$$

or

$$U = \sum_\alpha \frac{\varepsilon_\alpha \omega_\alpha \exp(-\beta \varepsilon_\alpha)}{F(\beta)} = -\frac{\partial}{\partial \beta} \log F(\beta).$$

For further discussion of the material in this section, see Fowler,[11] Mayer and Mayer,[15] and Schrödinger.[16]

8.4. Relation to Thermodynamics

With the above probability interpretation, U is the expected value of the internal energy of the system, and the immediate question is to identify other thermodynamic functions in terms of this probability picture. F is a function, $F(\beta, \varepsilon_1, \ldots)$, of β and the energy levels and

$$d \log F = -U d\beta - (\beta/F) \sum \omega_\alpha \exp(-\beta \varepsilon_\alpha)\, d\varepsilon_\alpha$$

$$= -U d\beta - \beta \sum (n_\alpha/N)\, d\varepsilon_\alpha.$$

Schrödinger interprets $\Sigma(n_\alpha/N)d\varepsilon_\alpha$ as the expected value of the work done on an individual system by external agencies to increase the average energy, i.e., $\Sigma\, (n_\alpha/N)\, d\varepsilon_\alpha = -dW$, where W is the work done by the system. Thus,

$$d \log F + U\, d\beta + \beta\, dU = \beta\, dU + \beta\, dW = \beta(dU + dW)$$

or

$$d(\log F + \beta U) = \beta \, dQ = \beta T \, dS,$$

where dQ is the heat flow into a system. This implies that for $\psi = \log F + \beta U$, Ψ is a function of S, $\Psi = \phi(S)$, and $d\phi/ds = \beta T$.

Thus, $\log F + \beta U = \phi(S)$, and we must determine $\phi(S)$. But in terms of systems, β is intensive and U and S are extensive functions. The usual discussions of statistical mechanics now show that $\log F$ is an extensive function by considering the case of a system obtained by combining two subsystems. Hence, Ψ and S are both extensive and consequently they are linearly related, i.e., $k\Psi = S$ and $\beta T = \phi' = 1/k$ or $\beta = 1/kT$. k is called the Boltzmann constant. We have, then,

$$k \log F = S - U/T.$$

What is of some interest is to interpret S in terms of the probability picture. If n_1, n_2, ... is the most likely set of values, then $v(n_1, n_2, ...)$ is the number of ways of obtaining these values. In the classical case, using Stirling's formula freely, we obtain, since $\Sigma \, n_\alpha = N$, $\Sigma \, \varepsilon_\alpha n_\alpha = NU$,

$$\log v = \log N! + \sum (n_\alpha \log \omega_\alpha - \log n_\alpha!)$$

$$= N \log N - N + \sum n_\alpha(\log \omega_\alpha - \log n_\alpha + 1)$$

$$= N \log N + \sum n_\alpha(\log \omega_\alpha - \log n_\alpha) = N \log N + \sum n_\alpha(-\alpha + \beta \varepsilon_\alpha)$$

$$= N \log N - N\alpha + \beta NU.$$

But $\alpha = \log A = \log N - \log F$ and thus

$$\log v = N(\log F + \beta U) = NS/k$$

or

$$S = k(\log F + \beta U) = (k \log v)/N = k \log (v^{1/N}).$$

S is therefore a measure of the number of possibilities for the system in the most probable case. Notice that the statistical argument yields a formula for F and that U and S are expressed in terms of this formula.

As an example, consider an "ideal gas" as a collection of particles that undergo only elastic collisions in a spatial region $\mathfrak{A}$ with volume V_a. We ignore gravity. For a system consisting only of a single particle, the phase space is six-dimensional with coordinates x^1, x^2, x^3, p_1, p_2, p_3, where (x^1, x^2, x^3) is a point of $\mathfrak{A}$ and $-\infty < p_i < \infty$. If R is a small region in phase space containing the point x^1, x^2, x^3, p_1, p_2, p_3 and with volume $dx^1 \, dx^2 \, dx^3 \, dp_1 \, dp_2 \, dp_3 = dV$, then we can take $\varepsilon = (1/2m)(p_1^2 + p_2^2 + p_3^2)$ and $\omega = dV/h^3$. Thus,

$$h^3 F = \sum \exp(-\beta\varepsilon_\alpha)h^3\omega_\alpha$$

$$= \left(\int\right)^6 \exp\left[(-\beta/2m)(p_1^2 + p_3^2 + p_3^2)\right] dx^1 \, dx^2 \, dx^3 \, dp^1 \, dp^2 \, dp^3$$

$$= V_a \left[\int_{-\infty}^{\infty} \exp(-\beta p^2/2m) \, dp\right]^3$$

$$= V_a(m/\beta)^{3/2} \left[\int_{-\infty}^{\infty} \exp(-u^2/2) \, du\right]^3$$

$$= V_a(2\pi m/\beta)^{3/2}.$$

Hence,

$$\log F = \log V_a - \tfrac{3}{2}\log \beta + C.$$

This is $\log F$ for a single particle. Since $\log F$ is extensive, the F_L for a gas consisting of L particles is given by

$$\log F_L = L \log F = L \log V_a - \tfrac{3}{2}L \log \beta + LC.$$

The terms in $d \log F_L$ have been interpreted at the beginning of this section as $d \log F_L = -U_L \, d\beta + \beta \, dW_L$ for W_L, the work done by the gas, i.e., $dW_L = P_L \, dV$. Since $\log F_L$ has been expressed as a function of β and V,

$$U_L = -\frac{\partial}{\partial\beta}(\log F_L) = \tfrac{3}{2}L/\beta; \qquad p_L = \frac{1}{\beta}\frac{\partial}{\partial V}(\log F_L) = L/\beta V$$

or $U_L = \tfrac{3}{2}kLT$ and $p_L V = kLT$. If L_0 is the number of molecules in a mole of gas, then $R = kL_0$. If μ is the mole fraction represented by the given gas sample, the last equation becomes the familiar $pV = \mu RT$. The formula for U indicates that the specific heat at constant volume is $\tfrac{3}{2}kL_0 = 3R$.

Thus, in the case of a "perfect gas" one has theoretically proven relations to use instead of empirical rules. This discussion is based on classical dynamics and a requirement that the ω_i corresponding to a mesh interval R_i represent a large number of possibilities. On the other hand, by making a correct choice of h, we obtain an absolute entropy as distinct from just a difference of entropy between two thermodynamic states. We also obtain an absolute energy U rather than simply energy differences.

For further discussion of the material in this section, see Fowler[11] and Schrödinger.[16]

8.5. The Fine Structure of Matter

The above procedure yields the Maxwell distribution for velocities. The more general classical statistical mechanics provided insight into kinetic processes in gases, chemical reactions, thermodynamics, and electromagnetic radiation. Thus, probability theory supplemented classical science to provide a much more extensive analysis of the behavior of matter. There was a corresponding more sophisticated set of mathematical procedures that added probability concepts to the previous applied analysis of geometry, integration, and partial differential equations.

The very completeness of the theoretical picture represented by this formulation soon indicated inconsistencies. Radiation equilibrium was shown to be inconsistent with a continuous distribution of energy values, and experimental agreement was obtained only by assuming discrete energy levels. The discovery of the electron and the Rutherford structure of the atom indicated major discrepancies in the classical theory of electromagnetism and the specific heat of metals. The wave nature of light had been accepted because of interference phenomena, but the photoemission of electrons also indicated a particle nature. Furthermore, the electrons that had been considered particles also exhibited interference effects as if they had a wave character.

The resolution of these difficulties, which led to quantum mechanics, is described in readily accessible literature (see Born,[2] van der Waerden[18]). This was an extraordinary intellectual achievement that began in the nineteenth century with an empirical mathematical description of the line spectra of certain chemical elements. In the twentieth century these results, relativity theory, and continuing experimentation structured a deepening understanding of atomic phenomena. Their understanding in turn yielded an integrated viewpoint of physics, chemistry, and the phenomena associated with matter.

The immediate objective was the interaction of matter and radiation, but intellectually the quantum mechanics should be considered as rectifying and completing the theoretical picture of the classical exact sciences. One major element of the previous situation was preserved. Macroscopic phenomena were represented as the probabilistic consequences of the action of a very large number of very small subsystems governed by chance. In general these subsystems are not independent in the probabilistic sense; for example, in solids they are spatially related in their actions. But more fundamentally, these subsystems cannot be individually identified, and a correct assignment of probability is obtained only by taking this into account. However, when this is done, notions of thermal equilibrium still have the same probabilistic character.

A new approach was required to describe the behavior of the micro-systems. The observable output was essentially probabilistic but quite different from that due to the "particle systems" of classical mechanics, where probability could be attached to distributions in phase space. In the early steps of the development of quantum mechanics, particle phase space was quantized, i.e., the continuous range was replaced by discrete levels corresponding to spectroscopic observations. But this in turn led to the use of the "state" of the subsystem as the fundamental concept in the mathematical theory. The state of the system has essentially a wave character, and the original particle description became a set of rules for determining the possibilities for states. The wave character of the state of a microsystem may be associated with a function defined on configuration space, i.e., the space of coordinates. Computational procedures were quite different and aimed at computing the probability of an event associated with the micro-system.

This development had a very significant interaction with mathematics. The early formulations were efforts to adjust the classical Hamilton–Jacobi mechanics, electrodynamics, and probability theory to fit spectroscopic phenomena. The transition to wave mechanics was facilitated by known mathematical methods to solve partial differential equations that had led to "eigenvalue" problems, and these solutions are still of considerable practical importance. But a satisfactory logical treatment of quantum mechanics required more mathematically. The perturbation problem required a more abstract formulation in terms of linear operators. As a consequence, the spectral theory of self-adjoint operators in Hilbert space was developed by von Neumann and others. Dirac's axiomatic treatment of quantum electro-dynamics incorporated a theory of the electron and positron based on representations of the Lorentz group. Lie group theory and the associated theory of Lie algebras had been developed in the nineteenth century in connection with the theory of differential equations. These now took on a much more abstract form, e.g., C^* algebras, which permitted theoretical speculation concerning ultimate physical structures.

Another mathematical development was that of distributions (see Guelfand and Chilov[12]). The explicit description of distribution theory is in terms of linear functionals on spaces of functions, but it also can be considered as a development of the theory of Fourier transforms. In recent times there has been a considerable reformulation of the analysis associated with partial differential equations in terms of distributions. This has been justified by the considerable extension of capability.

Quantum mechanics produced a complete and satisfactory description of chemistry in the molecular sense and for radiation within a tremendous range of energy (see Herzberg[14] and Heitler[13]). This had many technological consequences, especially in electronics, after the Second World War. There was also a considerable understanding of nuclear structure, which produced nuclear weapons and an explanation of solar energy.

For further discussion of the material in this section, see Born,[2] Condon and Shortley,[5] Dirac,[8] Guelfand and Chilov,[12] Heitler,[13] Herzberg,[14] and van der Waerden.[18]

8.6. Analysis

The history of analysis should be associated with a precise exposition of the subject itself, but some general comments may be desirable. Both Leibnitz and Newton stated the general principle of the integral calculus, which permits one to evaluate definite integrals by finding antiderivatives. Thus, one solves a differential equation instead of summing algebraically, the process that had been used in the seventeenth century.

The Euler equations in the calculus of variation and integral relations such as Stokes' theorem or Green's theorem lead to differential equations either ordinary or partial. The procedure first used to solve differential equations was either infinite series or separation of variables followed by a series solution of the resulting ordinary differential equation. The convergence of infinite series was the first aspect of a more rigorous basis for analysis in which an axiomatic description of the Euclidean line corresponded to the system of real numbers. The corresponding geometric description of complex numbers as the Euclidean plane led to the theory of the analytic functions of a complex variable.

Infinite series analysis yielded existence and uniqueness results "in the small," i.e., the domain of existence would be finite but subject to *ad hoc* estimates of size. To handle existence and uniqueness in the large, two essentially different but complimentary approaches were used. One was the introduction of "topological" notions in the point set sense and the corresponding analysis of continuous functions and sequences of such functions. The other approach was that of infinite dimensional spaces in which a linear partial differential operator can be considered as a linear transformation. Each of these approaches involved a considerable amount of interacting development. For example, these required the extension of integration in the Lebesgue sense or in the Lebesgue–Stieltjes–Radon–Nikodym sense.

The linear transformation theory of differential operators began with the investigation of linear differential equations with constant coefficients, and for these the use of Fourier transforms or Laplace transforms was particularly effective. The Sturm–Liouville theory indicated a beautiful analogy between a more general class of differential operators and symmetric matrices in regard to characteristic values and vectors. Fredholm presented a precise basis for these indications by his analysis of the inverse integral operator. This in turn led to the introduction of Hilbert space by Hilbert. The spectral theory of self-adjoint operators has been further developed by Riesz, von Neumann, Carleman, Friedrichs, Stone, Lorch, and others using modern abstract concepts.

The "spectral theory" is applicable to "normal" operators, which one can think of intuitively as operators in which the characteristic vectors are mutually orthogonal. In the more general case the operator can be analyzed as a linear transformation in the sense of Banach from one linear space to another. The mapping characteristics of such a transformation are given by certain theorems of Banach and yield precise uniqueness and existence results. If the two linear spaces are Hilbert spaces a further analysis such as that given by the spectral theory is possible.

Sophus Lie showed that the possibilities for continuous groups could be analyzed in terms of certain linear partial differential operators and an algebraic structure based on these. The notion of invariant measure on such groups has led to a theory of the representation of groups and algebras of linear operators on linear spaces. This last generalizes the notion of matrix algebras on finite dimensional linear spaces. Such algebras, for example, von Neumann algebras or W^* or C^* algebras, are considered to have a close conceptual relationship with quantum mechanics.

Modern discussions of the structure of linear partial differential operators usually involve the concept of a distribution. One describes a linear set of functions, the "base function," and the distributions are linear functionals on this base set, i.e., they are functions defined on this set that are linear. For example, if the base set Φ consists of functions ϕ, and f is such that the integral exists for all ϕ in Φ, then $I_f(\phi) = \int f\phi \, dx$ is a linear functional on Φ and hence a distribution. The power of this concept lies in the fact that the set of linear functionals for a given Φ is much more extensive than the well-formed integrals. Smoothness and integrability properties of the base set Φ can be associated with the corresponding properties of the distributions. Base sets can be chosen so that analytic properties for the distributions hold, as do certain symmetries under the Fourier transforms. Since functions correspond to a subset of the distributions for a given base set, the problem of solving a partial differential operator equation $Lu = v$ can be generalized to the case where u and v are distributions. This yields a very useful technique for linear partial differential operators using the distributions of the delta function and its derivatives.

Exercises

8.1. Describe the Pascal triangle and show that the construction process yields the binomial coefficients.

8.2. Show that the number of the ways in which k bins can contain n indistinguishable objects is $\binom{n+k-1}{k-1}$. [*Hint*: Consider $n+k-1$ positions in order on a line and locate "separators" at $k-1$ of these places and put objects in the remaining positions.]

8.3. Evaluate the coefficient of x^p in the expansion of

$$(x+x^2+x^3+x^4+x^5+x^6)^n$$

for $n \leqslant p \leqslant 6n$.

8.4. What are the different forms of Stirling's formula and how are they proven?

8.5. If x and y are independent random variables with distribution functions $f(x)$ and $g(y)$, respectively, show that the distribution function for $z = x + y$ is

$$h(z) = \int_{-\infty}^{\infty} f(z-u)g(u)\,du.$$

Show that if $F(r)$, $G(s)$, and $H(t)$ are the respective Fourier transforms of f, g, and h, then $H(t) = F(t)G(t)$.

8.6. Two players match coins of a fixed denomination until a player loses all his coins. Suppose A has 5 coins, B has 15 coins. What is the probability that B will win?

8.7. Todhunter[17] (p. 295) expresses a result of Bayes as a ratio of integrals and states, "Bayes does not use this notation: areas of curves, according to the fashion of his time, occur instead of integrals." What is the definition of the integral that Todhunter uses in 1865?

8.8. A game begins with n markers on the table. Each player takes one or two markers from the table, and the winner is the one who takes the last marker on the table. Show that one can always win, provided one can assure that one's opponent must make a choice from a number of markers that is a multiple of three.

8.9. For the ε_i associated with regions R_i in phase space, discuss the restrictions so that the total number of possible distributions $\{n_1, n_2, \ldots\}$ is finite.

8.10. Discuss the normalization process for the relative probabilities $v(n_1, n_2, \ldots)$. In the sum of the permitted v, factor out the maximum term. This is analogous to certain discussions of the central limit theorem.

8.11. Show that the entropy for a perfect gas contained in a volume V with molecules of mass m is

$$S = \mu R[\log V + \tfrac{3}{2}\log(2\pi m/\beta) + C]$$
$$= \mu R[\log VT^{3/2} + \tfrac{3}{2}\log(2\pi mk) + C],$$

where μ is the mole fraction, that is, the number of molecules, L, is μL_0, and C is a constant independent of the choice of m. Notice the increase in entropy corresponding to an expansion of the gas into a vacuum of equal volume through a small aperture. The kinetic energy of the molecules will remain the same in such an expansion.

8.12. Suppose that the N systems of Section 8.3 are indistinguishable but that only one system can be put in a given state. Then, in terms of the binomial coefficients,

$$v(n_1, n_2, \ldots) = \binom{\omega_1}{n_1}\binom{\omega_2}{n_2}\cdots.$$

Assume ω_i is large. The most probable n_i are given by the formula

$$n_i = \omega_i A \exp(-\beta\varepsilon_i)/[1 + A \exp(-\beta\varepsilon_i)],$$

where A and β are constants determined by $N = \Sigma\, n_i$, $NU = \Sigma\, \varepsilon_i n_i$. The last equation may be replaced by an assumption that β is known. In this case, discuss how to determine A.

8.13. If the N systems of the previous exercise are indistinguishable but any number can be put in a given state, then

$$v(n_1, n_2, \ldots) = \binom{n_1 + \omega_1 - 1}{n_1}\binom{n_2 + \omega_2 - 1}{n_2}\cdots.$$

Assume at least $\omega_i > 1$ and proceed as in the previous exercise.

8.14. Show that if $\Sigma\, x_i A_i = 0$ for every $\{x_1, x_2, \ldots\}$ for which $\Sigma\, x_i B_i = 0$ and $\Sigma\, x_i C_i = 0$, then there exist constants α and β such that $A_i = \alpha B_i + \beta C_i$ for every i. (The $\{x_1, x_2, \ldots\}$, are restricted to sequences in which only a finite number of terms are not zero.)

8.15. Suppose Ψ and S are functions of variables $x_1, \ldots, x_n$, with $n \geqslant 2$ and $d\Psi = f(x_1, \ldots, x_n)\,dS$. Show that Ψ is a function of S, $\Psi = \phi(S)$, with $f = d\phi/dS$. [*Hint*: Change variables to a new set $u_1, \ldots, u_n$ with $u_1 = S$.]

8.16. Discuss the extensive character of the function $\log F$ relative to systems.

8.17. In an elastic collision two particles interact so that the total momentum and the amount of kinetic energy associated with each direction in space are unchanged. Obtain the corresponding transformation in phase space and the determinant of this transformation.

8.18. Show that a mole of one perfect gas differs from that of another perfect gas only in density.

8.19. If the energy is only kinetic for a classical collection of systems and homogeneous in the p_i and there are m pairs $\{p_i, q_i\}$, then one can extract β from the p_i integral and

$$\log F = \log V_a - \frac{m}{2}\log \beta + C.$$

Find S and the p, V relation for an adiabatic change.

8.20. Define a Hilbert space, $\overline{\mathfrak{W}}_1$, whose elements are functions defined for $0 \leqslant x \leqslant 2\pi$ and whose inner product is

$$(f, g)_w = \int_0^{2\pi} (f\bar{g} + f'\bar{g}')\,dx.$$

How is completeness proven?

8.21. Show that the sequence $\{e^{inx}/[2(1 + n^2)]^{1/2}\}$ constitutes an orthonormal set for $\mathfrak{W}_1$. This set is incomplete, since it is orthogonal to $\rho(x) = \sinh(x - \pi)/(\sinh 2\pi)^{1/2}$. If this function is added, show that one has a complete orthonormal set in $\mathfrak{W}_1$.

8.22. If $f(x) \in \mathfrak{W}_1$, show that

$$(f, \rho) = [f(2\pi) - f(0)]\cosh \pi/[\sinh(2\pi)]^{1/2},$$

and if $f = (f, \rho)\rho + u$, then $u(2\pi) = u(0)$. A necessary and sufficient condition for f to be expressible in $\mathfrak{W}_1$ in terms of the orthogonal set $\{e^{inx}/[2(1 + n^2)]^{1/2}\}$ is that $f(2\pi) = f(0)$. What does this mean concerning the term-by-term differentiation of Fourier series?

8.23. Show that if $g(t, x)$ is such that for each x, $g_x(t) = g(t, x)$ is in $\mathfrak{W}_1$, and such that for every $f \in \mathfrak{W}_1$, $(f, g_x)_w = f(x)$, then

$$g(t, x) = \cosh t \cosh(2\pi - x)/\sinh 2\pi \qquad \text{for } t \leqslant x$$

and

$$g(t, x) = \cosh(2\pi - t)\cosh x/\sinh 2\pi \qquad \text{for } t > x.$$

Note also that

$$g(t, x) = \sinh(x - \pi)\sinh(t - \pi)/\sinh 2\pi + \sum_n e^{in(t - x)}/2\pi(1 + n^2).$$

The function $g(t, x)$ is analogous to a "reproducing kernel."

8.24. Let T be a transformation from $\mathfrak{W}_1$ into a space $\mathfrak{H}$. Consider the following two problems: (a) Determine N_1, the set of f values in $\mathfrak{W}_1$ such that $Tf=0$. (b) Given an $h \in \mathfrak{H}$, determine an f_0 such that $Tf_0 = h$. If T has an adjoint with dense domain, T^*, then $T^*: \mathfrak{H} \to \mathfrak{W}_1$, has a range R_R^* whose orthogonal complement is N. How can this be used to determine N?

8.25. A formal solution of Exercise 8.24(b) is obtained by finding for each value of y a function u_y in L_2 such that $T^*u_y = g(t, y)$. Then

$$(h, u_y) = (Tf_0, u_y) = (f_0, T^*u_y) = (f_0, g(t, y)) = f_0(y).$$

Discuss this formal procedure.

8.26. The results in the preceding two exercises were dependent on having T^* available. Let $\mathfrak{H} = L_2(0, 2\pi)$. Since $\mathfrak{W}_1 \subset L_2$ corresponding to T, there is a transformation, $S, L_2 \to L_2$. Thus, S is an operation on functions, and we assume S^* can be obtained. Let L be the transformation $\mathfrak{W}_1 \to L_2$ given by

$$Lf(x) = f(x) = (f, g_x)_w = \int_0^{2\pi} \left[f(t)\bar{g}(t, x) + f'(t)\bar{g}'(t, x) \right] dt.$$

Then

$$(Lf, h)_2 = (f, L^*h)_w \quad \text{for} \quad L^*h = \int_0^{2\pi} g(t, x)h(x)\, dx.$$

We have

$$T = SL, \quad T^* = L^*S^*,$$

and

$$T^*h = \int_0^{2\pi} g(t, x)S^*h(x)\, dx = \int_0^{2\pi} S_x^+ g(t, x)h(x)\, dx,$$

formally. When are these manipulations justified?

References

1. Blackwell, D., and Gershick, M. A., *Theory of Games and Statistical Decisions*, John-Wiley and Sons, Inc., New York (1954).
2. Born, Max, *Atomic Physics* (English translation), Hafner Publishing Company, New York (1956).
3. Bragg, Lawrence, X-ray crystallography, *Sci. Amer.* **219** (1), 58–70 (July 1968).
4. Burger, Ewald, *Introduction to the Theory of Games*, Prentice-Hall, Englewood Cliffs, New Jersey (1963).
5. Condon, E. U., and Shortley, G. H., *Theory of Atomic Spectra*, Cambridge University Press, Cambridge, England (1963).
6. Cramer, Harold, *The Elements of Probability Theory*, John Wiley and Sons, Inc., New York (1955).
7. Cramer, Harold, *Mathematical Methods of Statistics*, Princeton University Press, Princeton, New Jersey (1946).
8. Dirac, P. A. M., *The Principles of Quantum Mechanics*, Oxford University Press, Oxford, England (1935).

9. Feller, William, *An Introduction to Probability Theory and Its Applications*, John Wiley and Sons, Inc., New York (1950).

10. Flower, Raymond, and Jones, Michael Wynn, *Lloyds of London*, David and Charles, London (1974).

11. Fowler, R. H., *Statistical Mechanics*, Cambridge University Press, Cambridge, England (1966).

12. Guelfand, I. M., and Chilov, G. E., *Les Distributions*, Dunod, Paris (1962).

13. Heitler, W., *The Quantum Theory of Radiation* (3rd edition), Oxford University Press, Oxford, England (1957).

14. Herzberg, Gerhard, *Atomic Spectra and Atomic Structure*, Dover Publications, Inc., New York (1944).

15. Mayer, E. M., and Mayer, M. G., *Statistical Mechanics*, John Wiley and Sons, Inc., New York (1947).

16. Schrödinger, Erwin, *Statistical Thermodynamics*, Cambridge University Press, Cambridge, England (1952).

17. Todhunter, I., *A History of the Mathematical Theory of Probability*, MacMillan and Company, Cambridge and London (1865).

18. Van der Waerden, B. L., *Sources of Quantum Mechanics.*, Dover Publications, Inc., New York (1968).

19. von Neumann, J., and Morgenstern, O., *Theory of Games and Economic Behavior*, Princeton University Press, Princeton, New Jersey (1944).

9

The Paradox

9.1. Intellectual Ramifications

Mathematics has always been an essential technical element in civilization, but it also plays an intellectual role because of the unique character of its conceptual structure. Geometric and arithmetic ideas framed and supported Babylonian astrology and as a consequence indicated order and inevitableness in human affairs. Classical geometry presented the original format for a rigorous logical discussion and shaped the whole concept of a philosophical point of view based on a specific set of principles from which all others are deduced. Problem mathematics is necessarily associated with the idea of human affairs based on a mutual understanding arrived at by logical means.

The mathematics of the sixteenth and seventeenth centuries was more than a practical conjunction of geometry and problem mathematics. Involved also were philosophical ideas of infinity and motion. But in addition there were intellectual imperatives for unity and a need for alternatives to philosophical systems that were inadequate for the extraordinary expansion of experience that was occurring. The opening up of a world of mathematical ideas complemented the telescope, the microscope, and geographical exploration.

The natural philosophy of Newton offered an understanding of the solar system on a dynamic rather than purely geometric level and also opened up tremendous practical possibilities, for example, the use of machinery. This certainly altered notions of a homocentric universe. It resulted in a general conviction that everything we deal with could be explained by proper physical laws and mathematics. These physical laws were to be established empirically by experience in the form of experiments, and many such laws were discovered relative to the elastic properties of gases and solids and the behavior of electricity. The developments in the eighteenth and nineteenth centuries consisted precisely in an expansion of both the known areas that were subject to empirical natural laws and the mathematics required to provide an effective theory. The basic conviction that there exist completely scientific explanations gradually became dominant.

The mathematical nature of the theory was strongly inducive of efforts to establish unifying structures. This contributed to the growth and character of scientific theory. Mechanics was subjected to increasingly sophisticated formulas, general principles involving energy were applied to a wide range of phenomena, and chemistry developed both in terms of molecular structure and energy. Biology was also transformed from the classification concepts of Aristotle to a dynamic form in which the phenomenon of life is seen to be subject to an evolutionary principle stated in terms of probability and an explicit description of the mechanical and chemical behavior of the material associated with the living processes. Thus, conceptually the sciences were unified on a deeper theoretical and axiomatic level.

These developments were completely integral with a tremendous expansion of mathematics. However, in the closing decades of the nineteenth century, mathematics entered a new phase. Intuitive elements in the conceptual structure were replaced by purely "logical" elements of axiomatic set theory. This new formulation contained isomorphic representations of the previously available analysis, and this corresponded to a more satisfactory rigorous development. A wealth of mathematical structures became available as a basis for scientific theories. Modern mathematics is inventive and permits setting up wide-ranging alternatives, and the choice must be narrowed by experimentation. This dual role is of course different from the situation in "natural philosophy" where geometry and the real numbers as ratios associated with the Euclidean line were part of a general axiomatic formulation involving all scientific magnitudes.

Much of modern analysis is associated with the quantum mechanics. This provided a knowledge of the fine structure of matter, which was a proper supplement for the macroscopic description of the behavior of matter, electricity, and radiation. This then was the completion of the process of developing the scientific understanding of the environment and our usual experience. We have then a reasonably complete mathematical scientific theory of mechanics, chemistry, electricity, electromagnetism, and radiation.

Included in this scientific picture was an increasing understanding of the chemical processes in living matter. The electron microscope revealed a reasonably complete physical structure for the cell. But within this structure, the living processes appear to be long and complicated sequences of chemical reactions between extremely complex organic molecules. The shape and chemical constitution of these molecules have been explored by many techniques, providing information that combined with basic thermodynamics seems adequate to explain the chemical behavior both in the case of metabolism and the hereditary processes. It is true that the complexity of living phenomena prevents a complete detailed description at present, but the area covered is continuously increasing and the basic principles of "stereochemistry" appear to be adequate.

Thus, the mathematical description of experience, which began with the planetary tables of ancient Babylonia, expanded into a complete formulation that included even living phenomena. This was the objective of centuries of effort, which finally produced an overall theory that reached back a billion years to encompass the origin and evolution of life in a convincing probabilistic framework. The cosmic horizons widened far beyond the solar system and not only in space but in time, stretching back to the origin of the present universe fifteen billion years ago. Our capabilities limit the range of experience to which scientific understanding is applicable. But these capabilities are rapidly expanding and there is no hint in our present experience of any boundary that will not be crossed ultimately.

But this completeness also implies exclusiveness. Since scientific understanding encompasses all experience, all other understanding must be either derivative or illusory. Philosophy cannot dig deeper than an understanding of all experience. Philosophical "principles" such as the dialectic format of "thesis, antithesis, synthesis" must either be shown to be a consequence of science, possibly subject to limitations or admitted to be illusory.

Certainly all the old animistic explanations of nature were swept aside. The gods who lashed their steeds in the tempests and the angels who carried the planets like lamps across the night sky were just figures on a tapestry woven in the youth of the race to hide the black abyss of ignorance. In the theory of evolution one has a scientific depiction of life from its beginning as conglomerations of naturally produced molecules. This understanding has a precise mathematical description in terms of chemistry and probability and eliminates completely the concept of the intervention by a deity in successive steps.

Indeed, the self-integrity of this understanding precludes the existence of God. He could not have created the world, since it has always been. Furthermore, any changes are accounted for by the development itself. God cannot introduce any change into this mathematically prescribed world without destroying it. And if God has no significance for our experience, He does not exist in any philosophically reasonable sense.

For further discussion, of the material in this section, see Gatlin[1] and von Neumann.[6]

9.2. The Paradox

Thus science has reached its natural pinnacle, displacing philosophy and religion. The ultimate basis of government decisions is science or at least a claim to be scientific. Any proposed public policy is justified or opposed on scientific grounds, and morality is modified, gradually, by what is scientifically available. Scientific terms and formats are glibly used far in excess of any appreciation of the disciplined understanding they are supposed to represent.

But there is one difficulty. Our proof that God does not exist also shows that we don't either. For how can we interfere in the determinate scientific world if God can't? The concept of an experiment in an empirical science involves the idea of a directed observation in which the pattern of effect is preset by our actions. This is emphasized by the use of instruments. The point is that an experiment requires directed observation—an action on our part, not a casual observation triggered by circumstances. In many experiments our interference in the environment extends to isolating some of it from part of its past history so that a hypothesis can be tested. Thus, we have contravened the deterministic development in order to increase our understanding. Indeed, the whole point of understanding is to intervene in the environment. Nevertheless, empirical sciences produce a mathematically determinate picture that includes the behavior of the material in our body. The use of probability does not alter this situation or permit animistic intervention as is clearly evident in the theory of evolution. Transition probabilities lead to mathematical predictions with no animistic aspect, just as Newton's laws do. Any animistic element is inconsistent with a probability description, i.e., the latter would be false if such an element is present. (See Murray.[4])

Thus, we have two alternatives. We can assume that we are capable of introducing variations into the environment that are not mathematical consequences of the past history. Or we can assume that we are automatons following present patterns but subject to the illusion that we can intervene. We will call these, respectively, the "action hypothesis" and the "automaton hypothesis."

The "action hypothesis" is consistent with our usual intuitive interpretation of our relation with the environment. We consider ourselves to be entities that are recipients of a continuous stream of impulses from the environment, which we integrate in our mind into an awareness of the world. Much of the integrand is from the past, i.e., stored, and we believe that we can detach this awareness from the present and substitute stored elements at will. Thus, awareness can roam in our imagination as we direct. Most of us believe that we can use this imagination to make plans, and furthermore, that we can direct our bodies to take action based on these plans.

Our awareness presents us as an entity immersed in the environment. In this environment we are aware of a natural stream of activity that would occur if we did nothing. We expect this activity to be scientifically determinate if no human intervention occurs. However, we can and do intervene in this natural stream of activity to direct it more favorably for ourselves. The problem is how, if our bodies are part of an essentially determinate environment, can these independent actions occur. Thus, E. Schrödinger in *What Is Life?*[5] (p. 85) states,

> (i) My body functions as a pure mechanism according to the laws of nature.
> (ii) Yet I know by incontrovertible direct experience, that I am directing its motion.

There are serious difficulties in making this picture consistent. One basic physical principle requires that action always be associated with a configuration of energy—i.e., a ball rolls downhill or the planets move in the solar system—and there is the question of how such a configuration can be arbitrarily introduced, since such an introduction is itself an action. It is an obvious aspect of controlled action that it corresponds to a cascaded sequence of actions, each of which triggers the next, and the mutual relations involved here offer no difficulty. But the initiation of the sequence appears to contradict one or the other of our basic notions.

The action theory is also called, pejoratively, "dualism." Any acceptance of an "animistic explanation" is certainly contrary to the direction of scientific development over many centuries. No direct positive evidence other than our subjective impressions is available to support this view.

The other solution is to assume that our ability to interfere in the world is really an illusion. Our organic structure is the result of a long period of evolution in which patterns favorable to survival were impressed on organisms by natural selection. These include patterns of behavior and the capability, like a computer, to store a summary of past events and to process them. The computer analogy here is critical, since we know that the computer does function as part of the deterministic environment and we can apply the evolutionary concept of natural selection both to the "program" of the computer and to the apparatus for gathering and storing data. Thus, our supposed ability to interfere in the environment is really an illusion. Our behavior, which produces this illusion, consists of transforming the data of past experience into rules governing our present activity. It is believed that procedures corresponding to this description can be set up in a computer, i.e., "artificial intelligence," "adaptive programming."

There are technical and practical difficulties associated with the programming techniques mentioned. An effective automaton must interact with its environment by receiving impulses from it and reacting. The computer, which must make the appropriate tie-in between input and output, deals only with the symbols associated with information and action. The actual effective information for which a symbol is to be entered may correspond to a considerable abstraction from the received set of impulses. One hopes the desired symbol is a function of the set of impulses received, but determining this function, the problem of "pattern recognition," is frequently difficult. Similarly, the action required may have a simple symbolic representation that is well understood, but the action itself may have to be resolved into a complex combination of specific motions that must be individually realized.

The proponents of this theory write as if these difficulties are the only issues involved and use an anthropomorphic language, using terms such as "sensing" and "decision." For example, the computer process of computing a function with discrete values is a "decision." This automaton model is fundamentally inadequate, and there are obvious difficulties with this approach, which has received widespread and complacent acceptance. The automaton is scientifically determinate, yet the scientific theories on which it is based involve experimentation containing directed and controlled actions for the purpose of observation. If this control and direction is an illusion, then the scientific theories lack an experimental basis.

Obviously, in an automaton there is no need for awareness. A computer does not have awareness. Since the actions of the automaton are predetermined, our awareness is a completely ineffective adjunct. Our decisions are determined for us by the overall development of events. Thus, awareness represents a duality in which it has no meaning, since awareness can be effective only by controlling action. This dualism is even more objectionable than the preceding.

9.3. Final Comment

Trapped by political considerations into committing what he knew to be a grave injustice, Pontius Pilate asked, "What is truth?" This reflected a certain familiarity with philosophy as it was taught in the ancient universities. The academic world no longer explicitly asks, "What is truth?" or "What is God?" or "What is mathematics?" The professional developers of knowledge deal with more sophisticated concerns whose significance in each case is apparent only to a small circle of cognoscenti.

The years have witnessed the growth of this vast coral reef of knowledge, and professional advancement in the universities has become associated with this growth. But most students in the university consider the objective of their education as their personal development, and there are incompatible elements in these objectives. The student with facility in certain intellectual exercises is frequently enticed into a complete acceptance of the supremacy of knowledge. The academic profession recruits itself from these. This produces a subculture with extremely separatist tendencies. The reaction against this is perhaps even more unfortunate, the belief that action is incompatible with intellectual development.

This may be the most critical aspect of our culture. Underneath many of the tendencies of our times are the moving forces of this split. Men of power, desirous of action, have developed an impatience with intellectual understanding, and movements that appeal to the academically oriented, such as "saving the environment," easily assume the objective of stopping all action.

The student may find his opportunities rather unpleasantly restricted by this schism. Nevertheless there is available to him a great hoard of intellectual treasures, and his immediate concern should be not with such questions as "What is truth?" or "What is mathematics?" or even "What can I do?", but with "What can I do with my mind?"

Exercises

9.1. Consider the relationship of the information concept considered in the Gatlin[1] book and the "information" in a blueprint. The latter should yield a structure when properly interpreted. How is the "information of the living system" to be decoded and how does this relate the individual to his environment? Does the environment decode the message? How is this discussion related to the subject of the von Neumann[6] book? How is a living organism a "history of two billion years"?

9.2. Discuss the responses of Heisenberg,[2] Schrödinger,[5] and Whitehead (Joad[3], Chapter XX, introduces the ideas of Whitehead and gives further references) to the paradox of Section 9.2.

References

1. Gatlin, T. L., *Information Theory and the Living System*, Columbia University Press, New York (1972).
2. Heisenberg, Werner, *Physics and Philosophy*, Harper Brothers, New York (1962).
3. Joad, C. E. M., *Guide to Philosophy*, Dover Publications, Inc., New York (1957).
4. Murray, F. J., Mathematics and the exact sciences, *Phil. Math.*, **10**, 134–154 (1973).
5. Schrödinger, E., *What Is Life and Other Essays*, Doubleday, New York (1956).
6. von Neumann, John, *Theory of Self Reproducing Automa* (completed by Arthur W. Burks), University of Illinois Press, Urbana, Illinois (1966).

Index